Bill Jense~
9321 Kauffm~
South Gate,
LO 6-0273

D0554287

MATHEMATICS
IN EVERYDAY THINGS

MATHEMATICS
IN EVERYDAY THINGS

by

WILLIAM C. VERGARA

Author of *Science in Everyday Things*

Illustrated with 54 Diagrams

HARPER & BROTHERS, PUBLISHERS, NEW YORK

TO
PAT AND BOB

PREFACE

This book, as may be apparent to some, follows the form of the author's *Science in Everyday Things,* which appeared about two years ago. At that time, I must confess, I felt some doubts concerning the wisdom of presenting such a varied group of subjects in random question and answer form. But the warm reception of *Science* in this country and its subsequent translations abroad indicate a certain appeal for that type of presentation. Fortified by that thought, it seemed not too illogical to make a further effort on the general subject of mathematics and its relationship to the scientific principles hidden in everyday objects and occurrences. Matters of historical interest also have been included in order to show the development of mathematics and to help fix its place in our culture. There is a comprehensive Bibliography for the benefit of those who may want to do further reading on the subject. For that purpose I would recommend *Mathematics in Western Culture* by Morris Kline as an excellent starting point.

The general reader should not be intimidated by the mathematical tables given in the Appendix. They have been included merely on general principles since such tables are not usually found in the average person's library. If the reader decides to put mathematics to work, he will find them useful. If not, he can ignore them since their use will not be required by the text.

Mathematics in Everyday Things requires little of the reader in the way of a mathematical background. High-school students should have little difficulty in understanding most of what appears in these pages. But it does require a certain degree of intellectual curiosity and clarity of thought—characteristics that are becoming ever more important in this period of unparalleled scientific development. Science and technology are advancing at a rate that staggers the imagination. And, significantly, it's not a constant rate. It's more nearly analogous to a bobsled racing down a mountain side; each succeeding instant finds it accelerating at an ever faster rate. Where is this scientific landslide leading us? Perhaps a look into the past will provide a clue.

For thousands upon thousands of years man relied upon one major means of transportation: the horse. He bent his energies to the development of better and faster horses in order to increase the speed with which he could move about. He was so proud of his minor improvements that he would wager large sums of money that one particular horse could run slightly faster than another. How fast? Perhaps 40 miles per hour. Then, within an instant of time, relatively speaking, he produced vehicles that move over 600 times as fast! Apply this sort of reasoning to any branch of science and one conclusion is inescapable: the world has a scientific bull by the tail! Medicine, chemistry, electronics, geology, botany, physics, astronomy, and all the others are on the threshold of almost unimaginable break-throughs of knowledge.

But in spite of this rate of progress, most of us seem to have taken science pretty much for granted. This attitude, however, can't be ascribed to any reticence on the part of the scientific community. On the contrary, each morning the newspapers are sprinkled with the latest developments reported by science's many spokesmen, and, surprisingly enough, a significant fraction of the information is reasonably accurate. Furthermore, the rate of scientific progress has become such a standard subject for editorial writers and after-dinner speakers that the entire question, unfortunately, is in considerable danger of becoming a bore! Then why this complacency?

The *New York Times* summed up the underlying difficulty succinctly in an editorial of October 21, 1957. On that day there appeared, to my considerable astonishment, a scientific explanation, complete with mathematical formulas, of the so-called Doppler-shift method of measuring the distance to an earth satellite! This firsthand encounter with the scientific process must have considerably shaken the dignified *Times,* for it went on to say, "... it isn't just the sputnik that is up in the air; it is most of the rest of us. The scientist, even when doing his best to be simple, speaks another language.... Perhaps we had all better learn that language."

But to act on the *Times'* advice, we will have to know the nature of the language. Can it be a language of words—like French, German, or Spanish? Certainly not; if words were the language barrier, a glossary of technical terms would long since have become the uniting bond between scientist and layman. The language is not one of *words;* it is one of *thought.* The difficulty is not in *definitions;* it is in the manipulation of complex *ideas.* The solution lies not in the study of *semantics;* it lies in the methods of *logical reasoning.* In short, the language of science is mathematics. Joseph Fourier, the great French physicist, said of mathematics, "There cannot be a language more universal and more simple, more free from errors and from obscurities, that is to say, more worthy to express the invariable relations of natural things."

It may be a happy coincidence, we are told, that mathematics seems somehow to be related to the physical world. It may be a profound accident that permits us to predict natural occurrences on the basis of mathematical reasoning. But there is another point of view. The wonderful geometry of Euclid was once thought to be the only possible geometry. The 10 axioms were once thought to be *self-evident truths* because no thinking man could doubt their applicability to the physical world. We now know, however, that there are a great number of possible geometries—each as self-consistent as that of Euclid. If this is true, can there be any doubt that at least a few will *approximate* reality? The physicist and the astronomer do not seek perfection, after all; they hope only for practical utility in their theories. And in this thought lies one of the most important concepts of the art of scientific investigation—that of the physical model.

The laws of nature are obscure. Their effects are so numerous and so complex that it's usually hopeless to attempt an exact mathematical analysis of a physical phenomenon. Instead, the scientist visualizes a simplified model which approximates reality and he then proceeds to study that model mathematically. A most famous physical model was that of the atom by Niels Bohr, which consisted of electron "planets" moving in orbits around a nucleus. Although Bohr's model later had to be supplanted by a radically different one, it was of inestimable importance in the development of chemistry and physics. In the study of science, a theory is related *exactly* to a physical model, and the model is related *approximately* to the real world. When a model fails to account for new observational data, it must be modified or discarded in favor of a new one. In this way, the "exact" sciences give us successively closer *approximations* of reality.

At this point I must make one reservation. In discussing the scientific method, it may seem that great physical theories evolve in some mysterious way from deep mathematical reasoning. With few exceptions, this seems *not* to be the case. To quote Nobel Prize winner Peter J. W. Debye, ". . . our science is essentially an art which could not live without the occasional flash of genius in the mind of some sensitive man, who, alive to the smallest of indications, knows the truth before he has the proof." The debt of science to mathematics lies not in its power to deduce truths about nature but in its ability to organize ideas and principles into general laws and then to test the entire scientific structure for logical soundness. It's not conceivable to me that this result could be obtained by other means.

But aside from the scientific importance of mathematics, one finds the subject to have a great beauty of its own. Why does one so enjoy a mathematical demonstration that leads unerringly to a precise conclusion? What is there about mathematics that so captivates the imagination? It's the harmony of the different arguments, their mutual dependence and their combined results; it's everything that introduces order, everything that unifies and simplifies; in short, it's everything that helps to provide a clear comprehension of the whole as well as the parts. It's no coincidence that this is precisely the attraction that science holds for the scientist. All too often, writers tend to divorce the two—scientists regarding mathematics as one of a number of valuable tools; mathematicians regarding science as an unexpectedly pleasant but probably irrelevant by-product of their art. Without attempting to be dogmatic, it is the author's opinion that, without mathematics, the exact sciences would revert to philosophical abstractions; without science, mathematics would remain a profound plaything. I hasten to add, of course, that I have no argument with profound playthings—or with philosophical abstractions. But, however valuable they may be as intellectual exercises, such pursuits can hardly be expected to provide anything approaching the amount of knowledge that has been made available by the marriage of mathematics and the scientific method. It has been a happy marriage, and a productive one; it has made our civilization possible.

WILLIAM C. VERGARA

Towson, Maryland
June, 1959

MATHEMATICS
IN EVERYDAY THINGS

MATHEMATICS
IN EVERYDAY THINGS

Are the sun's rays really parallel?
On a sunny afternoon not long ago, I was seated at my desk busily engaged at something or other when I noticed a row of oval patches of light on the desk top. It occurred to me that there must be circular apertures somewhere in the drawn Venetian blinds. Since the desk top was not perpendicular to the rays, the light patches would be elliptical rather than circular. That reasoning would seem to account for the light patches; I tried to get back to work. But where in the world could the circular openings be? I turned around and stared at the blinds for some time— and there were no circular apertures! By this time, work was out of the question, so I held a piece of paper against one of the light patches and traced it back to the blind. Paradoxically, the apertures in the blind that were causing the spots of light were decidedly *rectangular* in shape! And yet the resulting patches of light were circular on a piece of paper held perpendicular to the rays at a distance from the aperture! Nature, it seems, is always full of surprises. For the patch of light is not merely a shadow in reverse, *it's an image of the sun!* The opening in the blind is the "lens" of a "pinhole" camera.

To get a quantitative picture of the situation, we must keep in mind the fact that the sun is a rather large body and not merely a point source of light. As shown in Figure 1, we see the

1

sun as a body that subtends a considerable angle at the observer's eye. The angle is equal to that formed by an object 1 unit long at a distance of 108 units from the observer. For such relatively small angles (about $\frac{1}{2}$ degree), the angle is equal to $\frac{1}{108}$ radian. Radian measure is particularly useful for such angles because it gives the ratio of the important distances directly, without recourse to trigonometric tables.

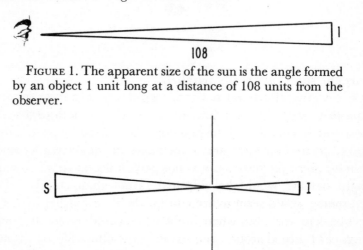

FIGURE 1. The apparent size of the sun is the angle formed by an object 1 unit long at a distance of 108 units from the observer.

FIGURE 2. An image of the sun is formed by rays of light passing through a small hole.

The second diagram (Figure 2) shows an image of the sun formed by a small hole in a piece of paper. It's apparent that the angle subtended by the image is equal to that subtended by the sun. The diameter of the image, then, must be $\frac{1}{108}$ times its distance from the aperture.

$$\text{Image diameter} = \frac{\text{distance from aperture}}{108}$$

If the aperture is small, a solar image formed 36 inches from the aperture will have a diameter equal to $\frac{36}{108} = \frac{1}{3}$ inch. Such an image will be faint but distinct. If the aperture is large, the im-

2

age will be bright and hazy, and its diameter will be somewhat larger.

You can produce a sharp image of the sun by piercing a small round hole in a sheet of paper or thin cardboard and holding it up to the light. It's even possible to see a large sunspot if one happens to be present at the time.

If you are still in doubt about these solar pictures, wait until a cloud passes over the sun; you will see it move across the patch of light—but in the opposite direction!

Concerning the parallelism of the sun's rays, it's evident that at least some of them depart considerably from our notion of parallel rays. The departure is so great that we can measure their angular separation with a single extension of an ordinary yardstick. To produce more nearly parallel rays we must select only those coming from a small area of the sun's surface. This can be done by placing two small apertures in line with the sun, thereby rejecting those rays coming from other than the selected area.

The story of sun pictures would hardly be complete without including the following passage: "I have . . . been much impressed by the courtesy of nature, which thousands of years ago arranged a means by which we might come to notice these (sun) spots, and through them discover things of greater consequence. For without any instruments, from any little hole through which sunlight passes, there emerges an image of the sun with its spots, and at a distance the image is projected upon any surface opposite the hole. It is true that these spots are not nearly as sharp as those seen through the telescope, but the majority of them may nevertheless be seen." The excerpt is from a letter written in 1612 by the versatile Galileo on the subject of sunspots. Very little indeed seems to have escaped his notice.

When does 1 plus 1 equal 1?

About one hundred years ago, August Ferdinand Möbius, the German astronomer, first investigated the curious properties of

ticians as one of the most profound and provocative of all of the branches of mathematics.

But to return to magic, for a moment, there is another topological trick that seems so impossible to the uninitiated that most of your friends will not even suspect it's a trick unless you betray the fact by your manner. Slightly dishonest gamblers have been known to subsist quite nicely with its help. To perform the trick you will need a belt about 3 or 4 feet long. To make the trick less obvious, try to use a belt that's identical at both ends. Then double the belt and twist it into a spiral as shown in the accom-

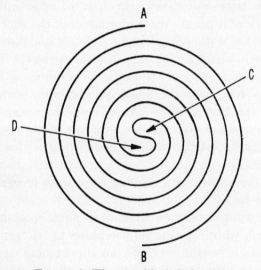

FIGURE 3. The gambler's belt trick.

panying diagram (Figure 3). Be sure to double the belt in such a way that the ends *A* and *B* are on opposite sides of the spiral at the conclusion of the winding. The unsuspecting spectator is then asked to place a pencil in one of the loops *C* or *D* so that when the ends are pulled away, the belt will catch on the pencil. It doesn't really matter which loop he selects, of course, as he will lose either way. The clever magician is able to control

5

the elusive loop and cause the belt to catch or pull free regardless of where the spectator places the pencil.

Here's how it's done. Suppose the spectator selects loop *D*, and you want him to lose. All you have to do is pull the ends away from the spiral in such a way that end *B* is *toward the outside*. Referring to the sketch, end *A* is allowed to drop down toward the left and the ends drawn quickly away. If loop *C* happens to have been selected, merely pull the ends away in the opposite manner; that is, push *B* upward and to the right so that end *A* is *on the outside* of the spiral as the ends are drawn away. Making one loop somewhat larger than the other will help you to keep track of it during the performance of the trick.

It always helps the effect if the spectators are allowed to follow the slowly turning loops as you form the spiral. This keeps their minds off what you are doing with the ends. When the spiral is complete, cover the outside portion of the spiral in order to hide the fact that the ends did not come out even. If the belt trick doesn't amaze your friends, you can surely suspect them either of prior knowledge or latent topological tendencies.

Why does the whistle of a moving train seem to change pitch as it travels by us?

We have all noticed the reduction in pitch of an automobile horn, or train whistle, as the vehicle passes by. In fact, the pitch is higher than it "ought" to be for an approaching sound source and lower than it "ought" to be for one that is receding. The effect of motion on the pitch of a sound was first explained in 1842 by the German physicist Christian Doppler who realized that a similar effect would take place for light waves. In 1848 the French physicist H. L. Fizeau applied the principle to astronomy in its current form.

To illustrate the principle, imagine a receding train with a whistle capable of emitting 600 vibrations per second. At a given instant, the whistle is turned on and the first vibration moves toward the observer. One second later the last vibration is produced and the whistle is turned off. Since the train was moving away from the observer during the 1-second interval, each successive vibration must travel a *slightly longer path* than its predecessor. For this reason, each succeeding vibration will be delayed by an amount of time depending upon the increased length of travel. If the observer measures the time interval between the reception of the first and last vibration, he finds it to be greater than 1 second by an amount that depends upon the speed of the train. As far as the observer is concerned, he hears a sound having less than 600 vibrations per second. In analagous fashion, the pitch of a similar approaching whistle is greater than 600 vibrations per second.

The general expression for the Doppler effect applies equally well to sound, light, and radio waves. That expression is

$$f = \frac{F(c - v_o)}{c - v_s}$$

where f = observed frequency, vibrations per second
  F = actual frequency, vibrations per second
  c = velocity of wave motion
   = 1,120 feet per second for sound
   = 186,000 miles per second for light and radio waves
  v_o = speed of observer away from source
  v_s = speed of source toward observer

The signs of v_o and v_s are chosen in such a way that both are positive in the same direction.

Imagine an automobile approaching you at a speed $v_s = 90$ feet per second (about 60 miles per hour). You happen to be in the middle of the street, so the driver's horn issues forth a continuous 400 vibration per second tone. What happens to the pitch of the tone as the car passes by? If you are essentially at rest, $v_o = 0$ and the expression becomes

$$f = \frac{400(1120 - 0)}{1120 - 90}$$

$$f = 435 \text{ vibrations per second}$$

7

You will *hear* a tone having a frequency of 435 vibrations per second as the car approaches. As the car recedes, the velocity v_s becomes negative and the expression becomes

$$f = \frac{400(1120 - 0)}{1120 - (-90)}$$

$$f = \frac{(400)(1120)}{1120 + 90}$$

$$f = 370 \text{ vibrations per second}$$

The ratio of the two frequencies is a measure of their effect on the ear. Since $\frac{435}{370} = 1.18$, the first tone (car approaching) will be 18 per cent higher in pitch than the second.

A half tone of the musical scale corresponds approximately to a ratio of 1.06 so the Doppler effect is responsible for the production of two audible sounds separated in pitch by 3 musical half tones.

Why are dice always made in the form of cubes?

The origin of dice is as obscure as that of playing cards, but it's interesting to speculate on the reasons for their cubic shape. From the earliest times, dice in Greece, Egypt, and the Orient were almost always constructed in the same manner as modern dice—with spots from 1 to 6 arranged on the faces of a cube. In addition, the spots on opposite sides always total the magical 7. But there's more than magic behind this arrangement. In order to insure equal odds for each face, each die will have to be a *regular polyhedron;* that is, all its faces will have to be identical. As a consequence of the work done by Leonhard Euler, it's known that there can be only five regular polyhedra. Of these, only the cube is easy to construct, and only the cube rolls just "right." The tetrahedron and octahedron have 4 and 8 faces respectively and do not roll well at all. The icosahedron (20 faces) and the dodecahedron (12 faces) are so nearly spherical that they quickly roll out of reach and are quite unsuitable even if they

8

weren't so difficult to construct. Since there just aren't any other regular polyhedra, the cube is the logical choice.

Since the sides of the cube are numbered from 1 to 6, it's the usual practice to arrange the spots so that each pair of opposite sides adds up to 7, and this principle is the basis of many tricks with dice. In one old trick, the magician turns his back and instructs an observer to throw three dice on the table. He is told to add the numbers turned up. He is then instructed to pick up any one die and add the number on the bottom to the previous total. This same die is rolled again and the new number that is turned up is also added to the total. The magician then turns around and after a suitable amount of thoughtful delay announces the correct total. To determine the total, the magician merely sums the numbers that are turned up; 7 added to this number gives the total obtained by the spectator.

What was the paradox of Achilles and the tortoise?

Long before the modern youngster is through grade school, he has learned to solve problems that baffled the greatest Greek mathematicians of antiquity. Calculation with pencil and paper was completely beyond their capability. So it's not too surprising to learn that the problems which tortured the most brilliant minds of the past fall easily before the mathematical methods that we have inherited. A typical example is the paradox of Achilles and the tortoise, one of a number of conundrums propounded by the philosopher Zeno. Achilles challenges the tortoise to a race. The tortoise is given 100 yards' head start. Achilles runs 10 times as fast as the tortoise. Here's what happens, says Zeno. Achilles runs 100 yards and reaches the point from which the tortoise started. But in the meantime, the tortoise has gone one-tenth as far as Achilles, and is still 10 yards ahead of him. Achilles then runs the 10 yards. Meanwhile, the tortoise has run one-tenth as far as Achilles and is 1 yard in front. Achilles then runs this 1 yard. Meanwhile the tortoise runs one-tenth of a yard, still maintaining the lead. Achilles runs this

9

tenth of a yard, but the tortoise runs a tenth of a tenth of a yard, and is still out in front by a nose. No matter how long the race continues, argues Zeno, the tortoise will continue to preserve his slim lead. Achilles is always getting closer to the tortoise, but he never quite catches up with him.

Of course, Zeno realized that Achilles *really did* win the race. What bothered him and his contemporaries was the failure of their methods of logical thought to solve the problem. "Where's the catch?" they wanted to know. Perhaps you're asking the same question. But as you will see, the use of elementary algebra, which was completely unknown to Zeno and Euclid, will solve the problem easily. Like Zeno, we know that Achilles will eventually catch up with and pass the tortoise. To solve the problem, we are really interested in finding the distance covered by the tortoise before Achilles takes the lead. Let's call that distance *x*, since it's unknown. We do know, however, that Achilles runs 10 times as fast as the tortoise. If the tortoise runs 1 yard in a certain length of time, Achilles will run 10 yards in the same length of time. The length of time that we select is not important, since it will be the same for both. So let's call their velocities 1 and 10 yards per minute, respectively. Now we can proceed to translate the problem from English into algebra for easier handling. We know that the distance a moving object travels is equal to its speed multiplied by the elapsed time. Or

$$\text{Distance} = \text{speed} \times \text{time}$$

Rearranging this equation, we can say that

$$\text{Time} = \frac{\text{distance}}{\text{speed}}$$

From the problem, we know that Achilles and the tortoise both will have been running for the same length of time when they come abreast of one another. So let's find the time that each runs, and equate the two. We find that the tortoise covers the distance, *x*, in $\frac{x}{1}$ minutes by substituting his speed (1 yard per

10

minute) and his distance traveled (x yards) in the equation given above. By the same reasoning, Achilles covers the longer distance, $x + 100$ yards (don't forget the head start) in $\dfrac{x + 100}{10}$ minutes. Since the two times are equal,

$$\frac{x}{1} = \frac{x+100}{10}$$
$$10x = x + 100$$
$$9x = 100$$
$$x = 11\tfrac{1}{9} \text{ yards}$$

The total distance that the tortoise runs before Achilles catches him is just $11\tfrac{1}{9}$ yards. While the tortoise is panting through his ordeal, Achilles sprints just $111\tfrac{1}{9}$ yards, no more, no less.

To understand why this problem presented such mathematical difficulties to Zeno and his friends, we must recall that such concepts as ratios and velocities were completely foreign to them. They could not do sums on paper. Any problem in long division was extremely difficult to them. They relied for all their calculation on the counting frame, or abacus. They knew that if a person adds progressively larger and larger quantities of anything, the pile grows with ever increasing speed as long as he persists in adding more. Why, then, they wondered, doesn't the same thing happen when *ever-decreasing* amounts are added indefinitely? It seemed to Zeno that we ought to be able to add smaller and smaller amounts to a pile without reaching a limit. In one case, the pile keeps increasing forever, at an ever faster rate, and in the other it goes on forever, at an ever slower rate. To understand their predicament, let's translate the problem into the kind of number system used by the Greeks and the Romans. Both used letters of the alphabet for numbers. Using such a number script, the sum of all the distances traveled by the tortoise before Achilles caught up with him would look like this

$$X + I + \frac{I}{M} + \frac{I}{C} + \frac{I}{X} + \cdots$$

11

(The three dots, which you will find often in this book, indicate that the series goes on forever.) Now let's put the same series of distances into our own decimal system.

$$10 + 1 + .1 + .01 + .001 + .0001 + .00001 + .000001$$
$$+ .0000001 + \cdots$$

Or, more simply,

$$11.1111111 \cdots$$

No matter how many 1's we put down, the number given above will never be as great as 11.2. Merely by the use of our modern number system we have come to the inescapable conclusion that it is possible to add ever-*decreasing* amounts to a pile *forever* without accumulating much of a pile at all! The Greeks and the Romans failed to discover this truth because of the clumsy system of numbers that they inherited. It was not until some unknown Hindu invented zero that modern mathematics became possible.

You will recall that $\frac{1}{9} = .1111111 \cdots$ so the number $11.1111111 \cdots$ turns out to be $11\frac{1}{9}$. The tortoise, therefore, is found to have run the same distance whether calculated by algebra or determined by the reasoning given above. Aside from solving Zeno's problem, our number system has demonstrated an extremely important mathematical principle. Mathematicians call it the convergence of an infinite series to a limiting value. In plain language, they mean that it's possible to add ever-decreasing quantities to a pile indefinitely until a point is reached at which the pile ceases to grow.

Are parallel lines really parallel?

About four thousand years ago, man began to get interested in counting and measuring things. It seems probable that much of the credit must be given to the Chinese for the early work done along these lines. Unfortunately, their clumsy system of picture writing seems to have prevented the Chinese from realizing their great early promise. It remained for the Greeks, some fifteen hundred years later, to produce an amazing amount of first-class mathematics in an incredibly short period of time. The two men that are usually given credit for founding Greek geometry were Thales (640–546 B.C.) and Pythagoras (582–507 B.C.). Paradoxically, both were of Phoenician parentage. Pythagoras traveled extensively in Egypt and India, where he picked up a great deal of knowledge of mathematics and the mystical nonsense that was associated with it in those days. Later, he founded a community in Croton, a Greek colony in southern Italy, dedicated to the study of mathematics and to the practice of mystical doctrines. In accordance with their Greek religion, the Pythagoreans believed that it was necessary to purify the soul from the taint of the physical and redeem it from its prison within the body. They wouldn't wear wool clothing, touch a white cock, use iron to stir a fire, or leave ashes on a pot. Members of the community were pledged to secrecy and were required to sign up for life. The penalty for disclosing important mathematical secrets to nonmembers was death.

In spite of their mystical doctrines, the Pythagoreans are credited with having been the first to apply deductive reasoning exclusively and systematically to the solution of mathematical problems. In line with their religious beliefs concerning the corruption of physical things, they completely divorced mathematics from practical considerations, and proved the fundamental theorems of plane and solid geometry and of the theory of numbers.

Another important school was the Academy of Plato, with Aristotle as its most influential student. Like the Pythagoreans, Plato's students emphasized "pure" mathematics, to the point of excluding all practical applications. In a similar manner, Greek mathematics sprang up all over the Mediterranean coast from Asia Minor to Sicily and southern Italy. The work of these mathematicians was later unified and simplified by Euclid in his great book, the *Elements*. Euclid started by selecting 10 well-chosen definitions, or *axioms,* that appear so obviously true that men have been willing to accept them for thousands of years as a basis for further reasoning. From these, he deduced some 500 theorems which constituted all the important results of his contemporaries. Typical axioms were "it shall be possible to draw a straight line joining any two points"; "it shall be possible to draw a circle with a given center and through a given point"; and: "the whole is greater than any of its parts." A typical theorem deduced from such axioms is "a diagonal of a parallelogram divides it into two equal triangles."

Perhaps you are wondering why the Greeks started off with geometry instead of something simple, like arithmetic. Why did Euclid struggle through a complex branch of mathematics at a time when the Greeks could work problems in arithmetic only with the help of their adding machine, the abacus? The answer lies in the needs of their peculiar culture. Five hundred years before Christ, the people of the Mediterranean area were engaged in surveying land, in building temples and palaces, in navigation, and in observing the heavens in order to tell time and maintain calendars. Such activities deal with the spacial properties of their environment. They had to work with lines, angles, triangles, and circles. Consequently, the body of mathematical knowledge that interested them concerned itself primarily with shapes and configurations, and only incidentally with numbers. Since the surfaces that Greek geometry dealt with were flat, their mathematics began with a study of plane surfaces.

Aside from its usefulness and beauty, Euclidean geometry provided an even more significant contribution to man in his rise from the intellectual depths—it showed him how to reason.*
Probably no other single creation of man has demonstrated how much knowledge can be derived by reasoning alone as has the geometry of Euclid. On the debit side of the ledger, the Greeks went a bit too far with their new-found technique. Deduction of so many useful and profound results from so few axioms led the Greeks and later civilizations to overestimate the power of rea-

* An opinion of the author which is not universally accepted in pedagogical circles, especially those untainted by an inordinate love of mathematics.

son alone. Theologians, scientists, philosophers, and all thinking men imitated the methods of Euclidean geometry. Aristotle, who probably did more than any other single man to hold back scientific progress, insisted that each science must consist of the deductive demonstration of truth from a few fundamental principles. As a result, the experimental and inductive methods of modern science were almost completely neglected for over a thousand years. The basic fallacy of the Greek philosophy was its assumption that the axioms were fundamental truths about nature—facts that were irrefutable. Modern scientists and mathematicians are much more cautious in this regard. Take parallel lines for example. Euclid defined parallel lines as straight lines that never meet, *no matter how far we go on drawing them.* The truth of the matter is, we just don't have any surface flat enough to allow us to go on drawing lines as far as we like and still keep them straight. Euclid, too, must have been concerned in this regard, for he never uses the parallel-line axiom until he has proved as many theorems as he can without it. In addition, he never assumes an infinite straight line to start with. When necessary, he extends a line in either direction only as far as the needs of the theorem require.

In the centuries that followed, many mathematicians concentrated on the parallel-line axiom and tried to deduce it from the other more reputable axioms, or to find a more acceptable substitute. But hundreds of attempts by the best mathematicians ended in failure. The parallel-line axiom stood against all assaults and became the scandal of geometry.

During the early part of the eighteenth century one man almost succeeded in solving the problem. The Jesuit priest Girolamo Saccheri, a professor of mathematics at the University of Pavia, came up with a new idea. He argued that the parallel-line axiom presented three possibilities. If we assume a line L and a point $P,$ then

   a. There is only *one line* parallel to L through P (Euclid's axiom).
   b. There are *no lines* parallel to L through P.
or c. There are at least two lines parallel to L through P.

His procedure was to take alternatives b and c successively, and develop two new geometries based on Euclid's other nine axioms and each of the new ones in turn. If such a procedure were to develop theorems that contradict each other, then he could surely infer that Euclid's geometry is truth and truth Euclid's geometry.

By using alternative b (no lines) Saccheri did deduce theorems that contradicted themselves, so that approach ended happily. But alternative c (two or more lines) *failed to produce any contradictions!* When added to Euclid's nine other axioms, alternative c produced a consistent geometry *with no contradictions!* Saccheri was on the threshhold of a fantastic discovery—but he couldn't discard a two-thousand-year-old habit of thought. The new theorems that he developed were so strange that he concluded that Euclid's parallel-line axiom must be right. In 1733, he published his conclusions in *Euclid Vindicated from All Defects.*

During the early nineteenth century, however, the mathematical climate was different and his work received critical re-examination. Three men, working independently, came to a different conclusion. Karl Gauss, Nikolai Lobachevsky, and János Bolyai followed Saccheri's line of thought and arrived at similarly strange theorems. But they refused to subvert their mathematical objectivity. They concluded that *there can be other geometries every bit as valid as Euclid's!*

Not long after, Bernhard Riemann (1826–1866), a young German mathematician, came up with a workable non-Euclidean geometry that seemed to make sense. First of all, he said that a line could be endless, but not infinite. In other words, he assumed that all lines, like the equator of the earth, are endless but finite. Secondly, he assumed that any two lines eventually meet. Finally, he assumed that two points may determine more than one line. Riemann then substituted his new axioms for the corresponding Euclidean axioms and developed a new set of theorems. Some of these are the same as Euclid's, since they depend only on axioms that are common to both geometries. But some of those that differ are striking indeed (Figure 4). For example: *All perpen-*

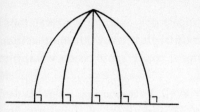

A

B

FIGURE 4. Riemann's geometry. (A) All perpendiculars to a straight line meet in a point. (B) Two straight lines enclose an area. (C) The sum of the angles in a triangle is greater than 180°.

C

diculars to a straight line meet in a point. Two straight lines enclose an area. The sum of the angles in a triangle is greater than 180°. Nonsense? Let's see how well these axioms and theorems fit the world we live in, before we reject them.

First let's look at the axioms and try to interpret them in terms of the sphere we live on. Let's assume the straight line of Riemann's axioms to be the shortest distance between two points *on the earth's surface.* Since the surface is spherical, such a "straight" line is nothing more than a *great circle*—that is, a circle whose center is at the center of the earth. Navigators have been using this fact for hundreds of years. In accordance with Riemann's description, great circles are endless but finite. In addition, great circles always intersect, so there can be no parallel lines on the sphere. His axiom about two points determining more than one

17

line is equally true, since some pairs of points on a sphere may have more than one great circle through them. A little reflection will show that Riemann's axioms *do* make sense when we apply them to the surface of the earth.

But how about the theorems? Well, let's examine a few. If the equator is a straight line, then all meridians of longitude are perpendicular to it. Yet all of these meet at a point, the North or South Pole! Similarly, since every pair of great circles must meet at two points, Riemann's straight lines must enclose an area. The same sort of reasoning will show that the sum of the angles in a triangle is greater than 180°. If we select two meridians of longitude and a piece of the equator, we can form one of Riemann's "triangles." True, this triangle may have "bent" sides in our Euclidean frame of reference, but such a triangle is certainly more useful to the navigator than one of Euclid's flat ones. Now the meridians of longitude make angles of 90° with the equator so these two angles alone add up to 180°. When we add in the angle formed at the North Pole, our sum must be greater than 180°.

The results of such investigations are inescapable—Riemann's geometry is a geometry of the physical world. It has mathematical significance and can be given meaning in an intuitively satisfying sense. But perhaps of greatest importance, non-Euclidean geometries such as Riemann's provided a climate in which the theory of relativity could develop. With unparalleled brilliance, Albert Einstein produced a geometry that revolutionized almost all branches of scientific and philosophic thought.

Just as the Pythagoreans had lifted mathematics from the taint of the physical, so has Einstein released man from the flat and sterile world of Euclid.

What are the basic laws of electricity?

In little over one hundred years, electricity has advanced from a laboratory curiosity to perhaps the single most useful element in our complicated age. And yet most of us take it quite for granted. What are the rules that govern this master servant of humanity? How does it work?

The story of electricity begins with the simplest of everyday things. Have you ever reached for the door handle of a car only to have your hand greeted by an unexpected burst of electricity? You can generate this kind of electricity by rubbing a glass rod or piece of plastic over various kinds of cloth. The object will then attract and hold small bits of paper. If you rub small inflated balloons over a piece of cloth, they will defy gravity by sticking to the walls and ceiling of a room. If the weather is dry, they will remain attached for weeks, or even months, at a time.

All these phenomena are examples of *static* (or stationary) electricity. When two objects are rubbed together, some of the electrons move from one object to the other. Following this operation, one object has a deficiency of electrons while the other has an excess. A neutral or *uncharged* object contains equal amounts of positive and negative electricity; when electrons are added to it, it becomes *negatively charged,* and when electrons are removed from it, it becomes *positively charged.*

The practical unit of electric charge is called the *coulomb.* A coulomb is the amount of electrical charge corresponding to 6.25×10^{18} electrons.

While static electricity may be interesting, it's not very useful. In order to put electricity to work, we must produce a movement of electrons. The motion of electrons along a piece of wire or any other electrical *conductor* is called an *electric current.* The magnitude of such a current is the rate at which electrons flow through the conductor—the number of electrons per second that move past a given point. The unit of electric current is, appropriately, the transfer of one coulomb of electric charge per second past some point in the conductor. Because the concept of current is so commonly used, a shorter name has been applied in place of *coulombs per second;* the unit is called the *ampere* after the French scientist André Ampère. If Q coulombs pass a point in t seconds, the current I at that point is

$$I = \frac{Q}{t} \quad \text{amperes}$$

Current, then, is merely the rate of flow of electrons.

In order to produce a current of electricity, we must have some sort of electrical force which can induce electrons to move. The battery produces such an *electromotive force*. The unit of electromotive force is the volt, named after the Italian physicist Alessandro Volta.

In 1827, the German physicist Georg S. Ohm first determined the quantitative relationship between current and voltage. He knew that a current would flow through a wire if its ends were attached to a source of voltage (electromotive force). Through experiments with such *circuits,* he discovered that the magnitude of the current is directly proportional to the voltage of the source. Doubling the voltage doubles the current. This relation is known as Ohm's law and is one of the most important in the entire study of electricity. It is expressed mathematically by the equation

$$I = \frac{E}{R}$$

where I is the current in amperes, E is the voltage in volts, and R is some number depending upon such characteristics as the length and diameter of the wire and the material of which it is made. The quantity symbolized by R is called *resistance*. It's somewhat analogous to friction in that it impedes the motion of electrons through the wire and is responsible for the conversion of electrical energy into heat. It's the resistance of the wires of an electric toaster, for example, that produces the desired heat.

The unit of resistance is the ohm. A wire, or any other electrical conductor, has a resistance of one ohm if a current of one ampere is produced through it by a voltage of one volt. If an ordinary electric lamp has a current of $\frac{1}{2}$ ampere flowing through

20

it when connected to a source of 120 volts, the resistance is equal to

$$R = \frac{E}{I}$$

$$R = \frac{120 \text{ volts}}{\frac{1}{2} \text{ ampere}}$$

$$R = 240 \text{ ohms}$$

If such a lamp operates for any length of time, it will get hot. The quantity of heat produced in a conductor depends upon the length of time the current flows and the magnitude of the current. The relation between heat energy and an electric current was determined empirically by James Joule

$$W = I^2 Rt$$

where W is the energy converted into heat in t seconds when a current of I amperes flows through a resistance of R ohms. The unit of energy is named the *joule* in honor of the discoverer of the law.

In the use of appliances, we are usually more interested in power, or the rate at which energy is expended. The unit of power measurement is the *watt*, which is equal to one joule per second. If P denotes the power in an electrical circuit,

$$P = \frac{W}{t} = \frac{\text{energy}}{\text{time}}$$

Since $W = I^2 Rt$, it follows that

$$\frac{W}{t} = I^2 R$$

and

$$P = I^2 R$$

21

In the example given earlier, the power used by the incandescent lamp is

$$P = I^2 R$$
$$P = (0.5 \text{ ampere})^2 \times (240 \text{ ohms})$$
$$P = 60 \text{ watts}$$

It was shown earlier that $I = \dfrac{E}{R}$ from which $R = \dfrac{E}{I}$, so

$$P = I^2 R$$
$$P = I^2 \times \dfrac{E}{I}$$
$$P = IE = \text{current} \times \text{voltage}$$

The equation $P = IE$ gives the power (in watts) as the product of the voltage (in volts) and the current (in amperes). Using this formula, the power used by the lamp is 120 volts $\times \frac{1}{2}$ ampere = 60 watts, which is identical to the power determined earlier.

Practical appliances, such as electric motors, are designed so that most of the electrical power is converted into useful work rather than heat. The *efficiency* of such a motor is defined as the ratio of the mechanical power delivered by the rotating shaft to the electrical power used up by the motor. The difference shows up as heat which raises the temperature of the motor. The power delivered by a motor is usually expressed in *horsepower*. Since one horsepower equals 746 watts, the formula for efficiency becomes

$$\text{Per cent efficiency} = 100 \times \frac{746 \times \text{horsepower output}}{\text{watts input}}$$

If a motor uses 1,000 watts of power and delivers 1 horsepower to a water pump, its efficiency is

$$100 \times \frac{746 \times 1}{1,000} = 74.6\%$$

22

How is a Moorish sundial made?

The first crude sundials, or shadow clocks, consisted merely of upright poles or obelisks mounted on a stone base. By measuring the angle of the sun's shadow with respect to fixed graduations (usually of 15° separation), it was possible to get a rough idea of the time of day. The trouble with such shadow clocks was their inaccuracy. The length of an hour as measured on the shadow clock varied with the season of the year when compared with the fixed interval of the hourglass, or waterclock. The Arabs, who made great headway in the study of spherical trigonometry, discovered that this behavior was a result of the changing angle of the sun as the earth progressed through its seasons.

The angle between the sun's shadow and the north-south meridian is defined as the sun's *azimuth*. But the azimuth angle—the angle through which the sun's shadow rotates when the sun has moved through a given angle—changes with the seasons. The Moors saw that this condition could be eliminated merely by aiming the shadow pole at the North Star, Polaris, instead of straight up in the sky (Figure 5). In other words, the axis of the shadow pole should lie parallel to the axis about which the sun appears

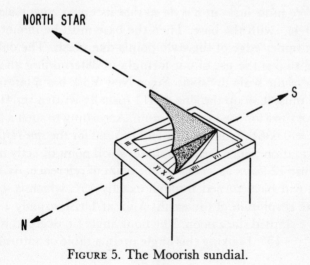

FIGURE 5. The Moorish sundial.

23

to rotate in its course through the sky. The proof of this discovery is beyond the scope of this book, but that will not prevent us from using the information, as the Moors did, to build an accurate sundial.

First of all, the style should be set so that it lies in a north-south direction. Then the top edge should be tilted above the horizontal at an angle equal to the latitude of the location for which it is made. The top edge of the sundial will then point at the North Star, since latitude is nothing more than the angle of the North Star above the horizon. When this has been done, the top edge will correspond to the earth's polar axis, and the scale can be graduated into divisions which will measure equal intervals of time at all seasons of the year.

The next step in constructing our sundial is to determine the angles for each of the hourly divisions on the scale. This can be done by using the following formula:

Tangent (shadow angle) = sine (latitude) × tangent (hour angle)

Here's how this formula is used. Let's suppose that we want to make a sundial for use in New York City, which has a latitude of 41°. We must first cut a style so that its upper edge makes an angle of 41° with the base. Then the base must be mounted so that the upper edge of the style points due north. The only remaining step is the use of our formula in determining an angle for each of the scale divisions. Since New York has a latitude of 41°, we must look up the sine of 41° (usually written sin 41°) in a table of sines for use in the formula. According to such a table, sin 41° = 0.6561. This number will be used for the *sine (latitude)* of the equation. Since the noon shadow will point directly north, we can use 12 NOON as a convenient point of reference. At 1 P.M. the sun will have turned through exactly 15°, which is $\frac{1}{24}$ of a complete revolution of the earth. Also, at 1 P.M. exactly 1 hour will have elapsed since noon. The hour angle (in degrees) is then $1 \times 15° = 15°$. Looking this angle up in a table of natural tan-

24

a strip of paper that has since been exploited by at least three generations of magicians. In the usual version, the magician hands a spectator three large paper bands, each formed by pasting together the ends of a long strip of paper. The spectator is then asked to cut the first strip in half lengthwise, cutting along the center of the strip until he returns to the starting point. As expected, the cutting results in two paper bands, each half as wide as the original. The unwary spectator is then asked to do the same with the second strip, and to everyone's amazement, the cutting results not in two separate bands, but in *one* narrow band that is twice as long as the original! Cutting the third strip produces an equally unexpected result—two paper rings that are interlocked!

All such tricks are based on the so-called *surface of Möbius.* The extraordinary surface is easily made by taking a long strip of paper and twisting it once before the two ends are joined together. When such a surface is cut completely around in a line parallel to the edge, only one band results. If the paper strip is given two twists before joining the ends together, two separate rings are produced by the cutting, but the rings are interlocked. Among the unusual properties of the surface of Möbius (one twist) is the fact that it has but one edge and one surface. Start any place on the surface and you can move completely around the ring to the starting point without crossing over an edge. The same property is true of an edge.

If you are wondering what all this has to do with mathematics, it turns out that the surface of Möbius is an example of a problem in *topology.* Although the word "geometry" may bring to mind a large number of theorems concerned with the measurements and relationships between distances and angles, it's a fact that some of the most fundamental properties of surfaces and volumes do not require any measurements of lengths and angles whatsoever. The branch of geometry that deals with such matters is *analysis situs,* or *topology.* It's regarded by many mathema-

gents, we find that tan 15° = 0.2679. The tangent of the shadow angle is then:

$$\text{tan (shadow angle)} = \sin 41° \times \tan 15°$$
$$= 0.6561 \times 0.2679$$
$$= 0.1758$$

Looking up 0.1758 in the table of tangents, we find that it falls between the angles of 9.9° and 10.0°, for which the tangents are 0.1745 and 0.1763, respectively. So we can take 10.0° as the shadow angle for 1 P.M. correct to a tenth of a degree. In calibrating the sundial, we would mark off an angle of 10.0° east of north for the mark corresponding to 1 P.M. Similarly, the angle corresponding to 2 P.M. would be determined by setting the hour angle equal to 2 × 15° = 30°, looking up tan 30° in the tangent tables, and substituting that number in the equation. In this way, all the hourly shadow angles can be computed and marked off on the scale.

After having completed your sundial, if you can induce it to ring at some predetermined time in the morning, you will have accomplished the first major improvement to the sundial in the past thousand years. Needless to say, those of us that are subject to the uncertainties of rural electrical systems would be eternally grateful.

Who first measured the size of the earth?

The first known measurement of the circumference of the earth was accomplished by the Greek mathematician Eratosthenes, who lived from 275 to 194 B.C. Eratosthenes was the librarian of the great library at Alexandria that had been established, along with a home for the Muses, by Ptolemy during the latter part of the fourth century B.C. In his capacity as librarian, Eratosthenes came across many interesting facts, one of which led directly to his measurement of the earth's circumference. He knew that at noon on a certain day of the year, the rays of the sun were reflected from the water in a deep well near Assouan, near the first cataract of the Nile. At this particular time, the sun is at its zenith and casts no shadow. At the same time at Alexandria, 500 miles to the north, Eratosthenes measured the length of the shadow cast by a pillar of known height. This showed the sun to be $7\frac{1}{2}°$ south from the vertical. With this elementary measurement, and a knowledge of four basic considerations, he was able to determine the earth's circumference. Here are the facts involved:

1. Rays of light coming from a distant source such as the sun appear to be parallel.

2. When a heavenly body is at its zenith, a line joining the heavenly body to the observer passes through the center of the earth.

3. When a line crosses two parallel lines, the corresponding angles are equal.

4. At noon, the sun is located directly over the observer's meridian of longitude.

Eratosthenes reasoned that two particular rays of the sun are of importance in this problem—the rays that pass through Assouan and Alexandria at noon on the day in question (Figure 6). By consideration 1 above, these rays are parallel. Consideration 2 makes it possible to extend one of these rays through Assouan to the center of the earth. He then drew a line up from Alexandria at $7\frac{1}{2}°$, the angle measured between the sun's rays and a vertical pillar. This line passes through the center of the earth because the pillar is vertical. Consideration 3 indicates that the angle at the center of the earth must also be $7\frac{1}{2}°$. And, finally, consideration 4 tells us that Assouan and Alexandria are on the same great circle, which is equivalent to the circumference of the earth. Eratosthenes knew the distance between the two cities to be 500 miles. This distance is that portion of the circumference corresponding to an angle of $7\frac{1}{2}°$. Since $7\frac{1}{2}°$ goes into a full circle of 360° approximately 50 times, the circumference of the earth must be $(500)(50) = 25,000$ miles.

In order to determine the radius of the earth, it was only necessary to use the value of π, (3.16), which the Egyptians had

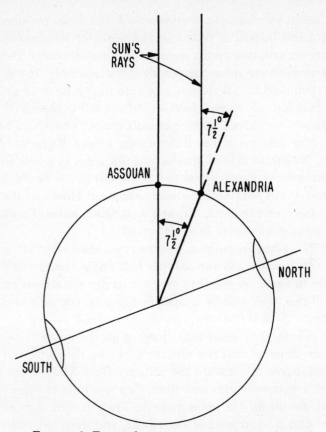

FIGURE 6. Eratosthenes measures the earth over two thousand years ago.

arrived at as early as 1500 B.C. Dividing the circumference by 2π gives the radius of the earth as approximately 3,950 miles. Man now knew that there was much more to the world than the portion that he was acquainted with. Some 1,700 years later, he was to begin doing something concrete about it.

What is the theory of relativity all about?

The physics of the nineteenth century was founded upon the principles of Euclidean geometry, and the ideas of absolute lengths, absolute time, and absolute simultaneity of events. A

27

1-foot ruler, for example, was exactly 1 foot long, no more, no less. A given instant of time was the same for all observers. If an observer saw two events happening simultaneously, then all observers must see them happening simultaneously. It was also firmly believed by all thinking people that a force of gravity exists. If it did not, then objects would not fall to the earth, and the planets would all whirl out into space. There can be no doubt—the universe acts as if there were a force of gravity. And, in fact, Newton's law of gravitation did seem to cover all the observed effects of gravity on the earth and in the heavens. *But the nature of this force has never been understood!* How can the sun exert a force on the earth through 93,000,000 miles of nothing? This question has never been answered.

In 1905, Einstein pointed out that two observers that are in motion with respect to one another will find it theoretically impossible to agree on whether two events are simultaneous. Because of this, they will be unable to agree on the time between two events. And, if two observers disagree about the simultaneity of two events, they must also disagree on the measurement of distances. Suppose that two observers, one on the earth, and one on Mars, agree to measure the distance from Mars to the sun. Since this distance varies with time, they will have to agree upon a given instant for the measurement. This, in turn, depends on the simultaneous striking of two clocks, and since two observers moving relative to each other cannot agree on the simultaneity of such occurrences, they will obtain different measurements for the distance from Mars to the sun.

In 1881, two American physicists, A. A. Michelson and E. W. Morley, had performed an experiment which indicated that the speed of light relative to the earth is not affected by the motion of the earth. Consequently, Einstein assumed as one of his axioms that the *velocity of light is independent of the motion of the observer.* Also, *no physical body has a speed greater than that of light.* And, *two observers at rest with respect to each other will agree on the distance and time between two events while observers in motion with respect to each other will not.* Einstein's theory indicates that such observers live in different time worlds. To illustrate, a man on a rocket ship traveling at about 160,000 miles per second would measure an object on the earth to be half its normal length. He would also note that a clock on the earth would move only half as fast as one on the rocket. An observer on the earth would notice the same discrepancies in length and time for objects on the rocket. Both sets of measurements are correct, but each is in terms of "local length and local time" in the space and time world of the observer.

Einstein disposed of the problem of the force of gravity by devising a new law that works without such a force. As a matter of fact, it seems to work a bit better, if anything. The planet Mercury had always failed to move around the sun in exact accordance with Newton's laws. But the new theory gives a path for Mercury that agrees, within experimental error, with the planet's observed motion.* Einstein's idea was as brilliant as it was shocking to scientists and philosophers. He proposed, essentially, a non-Euclidean geometry in which time is inextricably associated with space, and in which the shortest distance between two points is not the straight line of Euclid. After all, why do we feel that a force of gravity is essential to physics? Merely because we assume that objects move in the straight lines of Euclid unless acted upon by a force. Einstein's thought was to choose a formula for his new geometry such that the path of each planet around the sun is the resulting "straight line" in his geometry. In effect, this revises Newton's first law to read that a body undisturbed by forces will follow the curve that gives the shortest distance between two points—called a geodesic—in Einstein's new geometry of space and time. Einstein's world is a four-dimensional unity of space and time. He explains the observed motions of

* The relatively great eccentricity of Mercury's orbit results in correspondingly great changes in speed for different points in the orbit. This is in accordance with Kepler's law of equal areas swept out in equal lengths of time. Modern theory holds that Mercury's mass increases with increasing speed, a situation that would help to account for the very slight irregularities found in the planet's orbit.

objects in the universe in terms of the natural paths of bodies rather than through the invention of an artificial force of gravity.

What is the true interest rate charged on time-payment plans?

The true rate of interest charged on most installment purchase plans is just about twice as high as it seems. Suppose you purchase an item for $100 to be paid for in monthly installments over a 1-year period. The "credit charge" applied to the purchase is announced as 6 per cent of $100, or $6. But a little reflection will show that a $6 charge corresponds to a 6 per cent interest rate only if you have the use of $100 for 1 year. In practice, the unpaid balance is reduced more or less evenly over a 1-year period. This means that the average amount of the unpaid balance is about half the original amount, or $50. Specifically, suppose the item is purchased on the first of January and paid for in 12 monthly installments of $8.34. The amount of the loan—or the unpaid balance—is given below for each month.

Month	Unpaid balance
January	$100.00
February	91.66
March	83.32
April	74.98
May	66.64
June	58.30
July	49.96
August	41.62
September	33.28
October	24.94
November	16.60
December	8.26
	649.56

The average amount of the loan is determined by dividing the sum of the unpaid balances by 12. In this instance the average

30

amounts to $\frac{1}{12} \times \$649.56 = \$54.13$. The true interest rate is
$$\frac{\$6}{\$54.13} \times 100 = 11.1\%.$$

What are rational numbers?

Early in the study of number theory, it becomes evident that numbers fall into several categories. First comes the sequence of *natural numbers* referred to often in the expression, "as simple as one, two, three." Notice that zero and the negative numbers are studiously avoided. No one would think of saying, "as simple as zero, one, two," or "as simple as minus one, minus two, minus three." Quite to the contrary, the acceptance of zero and the negative integers as *bona fide* members of the general *integer field* was a milestone in the history of our culture—a milestone that was reached only through the expenditure of great effort and considerable pain.

In medieval times, most algebraists would have agreed that the equation $x + 7 = 3$ is impossible to solve since no number when added to 7 can produce a smaller number, 3. But since the human mind loves to generalize, equations of the type $x + a = b$ soon came into use in order to provide a *model* for all such equations. The indicated solution of the equation is $x = b - a$. Having written this solution, the impossible has been made possible—the meaningless has been given meaning. If a happens to be greater than b, the negative solution must indicate a *direction* or sense opposite to that normally considered positive. So negative numbers are indeed real numbers. Zero, far from being a great void, is merely the steppingstone between the negative and positive numbers, the connecting link that makes the operation of subtraction omnipossible.

Even with this enlarged integer field, division still presents a problem. If the division of any integer by any other integer is to be permitted, the positive and negative fractions must be included. This combination—the integers and fractions, positive

31

and negative, and the number zero—all these constitute the *rational domain*. Through them the four basic operations of arithmetic are made possible for all numbers with but one reservation, division by zero, which is discussed in another question.

What is the principle behind the bridge player's rule of eleven?

The most common opening lead at a no-trump contract is conventionally the fourth highest card in the suit led. If the player having the lead holds Q–9–8–6 in the desired suit he will lead the 6, his fourth highest card. The *rule of eleven* states that the number of cards in the other three hands that can beat this card are equal to $11 - 6 = 5$. If the 6 is led, there are five higher cards distributed among the dummy and the other closed hands.

Let's see how the principle works in practice:

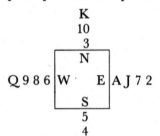

West begins the play by leading the 6—his fourth highest card. North follows with the 3. East subtracts 6 from 11 and determines that there are five cards distributed between the North, East, and South hands that can beat the 6. But East has three of them and the dummy two, so South must have none. East confidently plays the deuce, knowing that partner will win the trick just as surely as if he had peeked at South's hand. On the next lead, the entire suit is won by leading through dummy's king. The play of the jack at trick 1, on the other hand, insures dummy's making a trick in the suit.

Perhaps you have already discovered the principle involved. The cards of one suit can be arranged in order of rank as follows: 2–3–4–5–6–7–8–9–10–11–12–13–14 where 11–12–13–14 are J–Q–K–A, respectively. If a card of rank n is the opening lead, there are $14 - n$ cards higher than n in all four hands. But three of these are in the hand of the player making the lead. The other three hands, therefore, must hold $14 - n - 3 = 11 - n$ cards higher than n—hence, the rule of eleven. Notice that if the ace were ranked below the deuce we would have a rule of ten instead.

What is the apparent diameter of the sun when viewed from the other planets?

Most of us are familiar with direct and inverse proportions whether or not we bother to think about them as such. If ribbon costs 10 cents a yard, then 5 yards will cost 50 cents. As the number of yards goes up, the cost increases in *direct proportion*. In mathematical terms,

$$\text{Cost} = K \times \text{length}$$

where K is a *constant* depending on the units used for cost and length. In the ribbon problem $K = 10$ cents per yard.

Our use of inverse proportions in everyday life is somewhat subtler but quite commonplace, none the less. Take your car radio for example. As you travel away from a radio station on a long trip the signal strength gets weaker and weaker until, finally, you must search the dial for a stronger signal. The *greater* your distance from the station, the *weaker* the signal. Similarly, the apparent length of an object diminishes as it moves away from the observer. Mathematically speaking, the apparent length is *inversely proportional* to the distance:

$$\text{Apparent length} = \frac{K}{\text{distance}}$$

where K is a constant depending upon the units used for the other quantities.

In the problem at hand we are interested in comparing the apparent width of the sun W_A as seen from any planet with its

33

apparent width W as seen from the earth. Using the expression given above,

$$W = \frac{K}{D}$$

and
$$W_A = \frac{K}{D_A}$$

where D = distance from earth to sun
D_A = distance from a planet to sun

Dividing the second equation by the first gives the desired relationship -

$$\frac{W_A}{W} = \frac{D}{D_A}$$

or
$$W_A = (W)\frac{D}{D_A}$$

The following list gives the distances to the planets using the earth's distance as a unit. Mercury's distance, for example, is 0.3871 times the earth's distance.

Relative distance from planet to sun

Mercury 0.3871
Venus 0.7233
Earth. 1.0000
Mars 1.524
Jupiter 5.203
Saturn. 9.539
Uranus 19.19
Neptune 30.07
Pluto 39.46

The apparent width of the sun as seen from Mercury is $W_A = W\frac{1}{0.3871} = 2.6W$ (approximately), or about 2.6 times as great as from the earth. The apparent widths of the sun from the other planets can be found in the same way.

Planet	Apparent width of the sun
Mercury	2.6
Venus	1.4
Earth	1.0
Mars	0.66
Jupiter	0.19
Saturn	0.10
Uranus	0.052
Neptune	0.033
Pluto	0.025

Pluto, the most recently discovered planet, is so far distant that the solar disk is only one-fortieth as wide as when viewed from the earth. To get an idea of how really small this is, walk about three yards from this book and look at any period on this page (if your vision is that acute!). That's the apparent size of the sun when viewed from Pluto! The illumination of Pluto by the diminutive sun is 1,600 times *less intense* than that of the earth. This is because illumination obeys an *inverse-square* (distance) law $-\left(\dfrac{1}{40}\right)^2 = \dfrac{1}{1,600}$. And yet, the sun is so powerful that it would take about 150 of our full moons to equal the solar light falling on Pluto. Small though it may be, Pluto's sun is by far the brightest star in its sky.

Are even numbers related to the primes?

A deceptively simple theorem, known as the Goldbach conjecture, states that *every even number is the sum of two primes.* Consider the first few prime numbers, for example: 1, 2, 3, 5, 7, 11, 13, etc.

$$1 + 1 = 2$$
$$1 + 3 = 4$$
$$1 + 5 = 6$$
$$1 + 7 = 8$$
$$3 + 7 = 10$$
$$5 + 7 = 12$$
$$3 + 11 = 14$$
$$5 + 11 = 16$$

Rather simple, isn't it? Yet, in spite of the great effort expended on this theorem since its introduction in 1742, no one has been able to prove the statement conclusively, or find an example that would disprove it.

The first constructive step in that direction was taken in 1931 when Schnirelman, a Russian mathematician, succeeded in demonstrating that each even number is the sum of *not more than 300,000 primes!* Luckily, Vinogradoff, his countryman, was able to reduce this number to the sum of four primes. But the last two steps, from four to two, seem to be the most difficult of all. No one can tell how long it will take for a proof of Goldbach's conjecture to turn up—if, indeed, a proof is possible. In mathematics, it seems, the simplest things are often the most difficult!

Find the length of the chord *AB*

This puzzle involves three identical tangent circles with their centers on a line *GD* (Figure 7). Another line *GC* is drawn tangent to the third circle. The point of the puzzle is to find the length of the chord from *A* to *B*.

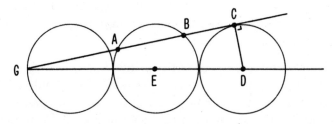

FIGURE 7. What is the length of the line *AB* in terms of the radius of the circle, *CD*?

If you want to try solving it, don't read on, for the answer follows. First of all, draw the perpendicular from point *D* to the line *GC* (Figure 8). Then draw another perpendicular from *E*, the center of the middle circle, to the line *GC*. Call the point of intersection with the line *F*. The triangles *GCD* and *GFE* are now seen to be similar since all three angles are equal. Then

$$\frac{CD}{FE} = \frac{GD}{GE} \qquad (1)$$

Now let the radii of the circles equal 5. (Any other number would do, but 5 works out nicely, as you will see.) The distance GD, being 5 radii in length will then be 25, the distance GE, 15, and the distance CD, 5. Substituting these numbers in (1) gives

$$FE = \frac{CD \times GE}{GD} = \frac{5 \times 15}{25} = 3 \qquad (2)$$

This tells us that $FE = 3$. But $EB = 5$, since it is a radius of the circle. Also, we know that triangle FEB is a right triangle. Now, a right triangle having a hypotenuse of 5 and a side of 3 must be a 3:4:5 triangle, so the other side, FB must be equal to 4. By similar reasoning $FA = 4$, so the chord $AB = 8$, or $\frac{8}{5}$ times the radius of the circle.

How is the age of an ancient object determined?

Most of us have walked through a museum and stopped to admire a relic of some long-dead civilization—an Aztec calendar, an Egyptian mummy case, or a war club of the ancient past. Our curiosity may take many diverse paths, but there is always one question in common, "How old is it?" Considering the difficulties involved, archaeologists have done quite well in determining the age of such objects. By comparing the writing on an object with historical records of the time, and by checking and cross-checking this information with other objects from the same geographical area, they have been able to tell us that a certain object is *about* 3,000 years old. But this just wasn't accurate enough to satisfy these men of science. So they began looking for a way of obtaining absolute dates for ancient objects in order to fill in the blank spaces of history.

The man who was to provide the solution to this problem was Dr. Willard F. Libby, of the Institute for Nuclear Studies at the University of Chicago—the institute directed by Dr. Harold Urey, the famous atomic scientist. Dr. Libby knew that there

37

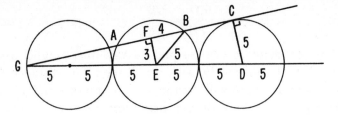

FIGURE 8. Solution to the puzzle of the tangent circles.

were two kinds of carbon, the ordinary kind which has a weight of 12, and a radioactive *isotope* which has a weight of 14. Both isotopes of carbon have the same chemical properties, but C-14 gives off rays over a period of time and disintegrates. He also knew that most living things on earth take up a certain amount of C-14 during their lifetime. This made it possible to determine the ratio of C-12 to C-14 to be expected in various living things. But when a plant dies, its atoms of C-14 begin to disintegrate, whereas the normal C-12 atoms do not change in quantity. The important point here is the extremely *slow rate* at which C-14 atoms disappear. This rate is so slow, so definite, that it takes many thousands of years before the atoms are completely gone. How does this work out in practice? Suppose that measurements show a certain tree to have 1 ounce of C-14 when freshly cut. One-half of this amount will decay during the first 5,500 years. During the next 5,500 years one-half of the remainder will decay. And during each succeeding 5,500 years the amount of C-14 present will be cut in half. C-14 is said to have a half-life of 5,500 years. As you can see, by measuring the ratio of C-12 to C-14 in an organic substance, and comparing this figure with the ratio found in living substances today, it is possible to determine an object's age accurately. Scientists have used this method on objects that are 20,000 years old! C-14 is the key which unlocks the secrets of the past and answers our question: How old is it?

If you're wondering how accurate the C-14 test really is, Dr. Libby performed some tests on an Egyptian funeral boat of

38

known age. Archaeologists knew with certainty that King Sesostris III died in 1849 B.C. Using the C-14 disintegration test, it was found that the age of the wood in the funeral boat taken from his tomb agreed very well with its known age. The C-14 test was mathematically correct.

One wonders what exciting information will be unearthed with this new scientific tool. History is being verified and in some cases rewritten through the use of C-14. This substance has become a sort of radioactive link between the past and the future—enabling us to learn about the past on the one hand, and performing work on other projects that will benefit all of us in the future.

Where can the midnight sun be seen?

Like a top spinning at an angle with the floor, the earth's axis of rotation is inclined $66\frac{1}{2}°$ to the plane of its orbit (Figure 9).

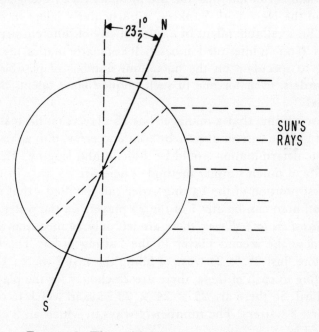

FIGURE 9. The midnight sun reaches over the North Pole in summer.

That plane, of course, includes the center of the sun. So, during the northern summer, sunlight reaches over the North Pole to illuminate a portion of the hemisphere that is on the nighttime side of the earth.

On June 22, the sun is at its most northern point in the sky so that its rays will reach $23\frac{1}{2}°$ over the North Pole. On that day the sun never sets above $66\frac{1}{2}°$ north latitude ($90° - 23\frac{1}{2}° = 66\frac{1}{2}°$). If we take into account the size of the sun's disk and the bending or *refraction* of sunlight through the atmosphere, the midnight sun can be seen as far south as $65\frac{3}{4}°$ north. Under such conditions, the sun is said to be *circumpolar*. The same phenomenon takes place in far southern latitudes during the opposite season of the year.

How many different baseball line-ups are possible?

Anyone familiar with major-league baseball is aware of the great number of line-ups that are possible. Casey Stengel, manager of the New York Yankees, is particularly adept in assembling his available talent in a multiplicity of different arrangements. If one is interested in baseball *and* mathematics, it's interesting to speculate on the maximum number of possible batting orders, even for one of such unquestioned talents as Mr. Stengel.

If we assume that a manager has 25 players on his team, we might get the correct answer by trial and error, but, as we shall see, the determination would be difficult and lengthy. There is, happily, a much simpler method. There are 25 ways in which the first position of the batting order can be filled—that is, the lead-off man can be any 1 of the 25 players on the roster. *Having selected the lead-off man,* there are left only 24 men from which to choose the second player in the batting order. There are, therefore, just 25×24 ways of filling the first 2 spaces. Corresponding to each of these, there are 23 choices for the player to bat third. So there are $25 \times 24 \times 23$ ways in which to choose the first 3 batters. The number of ways in which all 9 can be selected is

40

$$25 \times 24 \times 23 \times 22 \times 21 \times 20 \times 19 \times 18 \times 17$$

which amounts to a staggering 741,354,768,000 different batting orders.

If you object to including more than 1 pitcher in a single line-up, we can assume that 10 of the 25 are pitchers and that 1 of the 10 will always bat last in the line-up. The number of ways in which the first 8 positions can be filled is

$$15 \times 14 \times 13 \times 12 \times 11 \times 10 \times 9 \times 8$$

or 259,459,200. Since any 1 of the 10 pitchers can be used with any of the above arrangements, the total number of batting orders is 2,594,592,000.

Arrangements such as batting orders are known technically as *permutations*. A permutation is an arrangement or sequence of a group of objects. Take the letters *ABC*, for example. There are 6 permutations of these letters: *ABC, ACB, BAC, BCA, CAB,* and *CBA*. If a set of objects has 6 members, to take another example, the number of permutations is $6 \times 5 \times 4 \times 3 \times 2 \times 1 = 720$. In general, the number of permutations of n objects is

$$n \times (n - 1) \times (n - 2) \times \cdots \times 4 \times 3 \times 2 \times 1$$

Products of this kind occur quite often in mathematics and a special symbol has been devised to cope with them:

$$n! = 1 \times 2 \times 3 \times 4 \times 5 \times \cdots \times (n - 1) \times n$$

The term $n!$ is read *n factorial*. For example, $4! = 1 \times 2 \times 3 \times 4 = 24$; $7! = 1 \times 2 \times 3 \times 4 \times 5 \times 6 \times 7 = 5{,}040$, and so forth. With the help of this symbol, the formula for the number of permutations of n objects can be written

$$_nP_n = n!$$

where $_nP_n$ means "the number of permutations of n objects taken n at a time."

41

In the study of batting orders, you will recall, we were not interested in finding the number of permutations of 25 men taken 25 at a time, but rather, the number of permutations of 25 men taken 9 at a time. In general, the formula for n objects taken r at a time is

$$_nP_r = \frac{n!}{(n-r)!}$$

In the baseball problem, $n! = 25!$ and $(n-r)! = (25-9)! = 16!$; so

$$_{25}P_9 = \frac{25!}{16!}$$

After the appropriate factors are canceled out of the above expression, it reduces to precisely the same number that was determined earlier.

Before leaving this subject, we really should discuss that type of problem in which the sequence of a set of objects is unimportant. Consider the poker hand consisting of five spades, A-Q-7-4-3, for example. A player receiving such a hand would be equally pleased to receive the cards in reverse order, 3-4-7-Q-A, or in *any sequence* whatsoever.* In such situations we are interested only in the membership of the set and not in the sequence. The word *combination* is used to cover such sets. The number of combinations of n things taken r at a time is designated by $_nC_r$ and the appropriate formula is

$$_nC_r = \frac{n!}{(n-r)!(r!)}$$

The use of the formula for combinations can best be illustrated by solving a typical problem. Suppose 6 persons are willing to play a few rubbers of bridge. In how many ways can 4 persons be chosen from this group? To solve the problem we must find the number of combinations of 6 objects taken 4 at a time.

* Except, of course, in stud poker.

$$_6C_4 = \frac{6!}{(6-4)! \times (4!)}$$

$$= \frac{6!}{2! \times 4!}$$

$$= \frac{6 \times 5 \times 4 \times 3 \times 2 \times 1}{2 \times 1 \times 4 \times 3 \times 2 \times 1}$$

$$= \frac{6 \times 5}{2 \times 1}$$

$$= 15$$

The formula given above can also be used to determine the probability of getting any specified hand at a game of cards. Suppose it's desired to determine the chances of getting a full house at poker; that is, a pair of one rank, and a triple of another rank. The problem amounts to finding the number of favorable combinations (of a pair with 3 of a kind) and dividing this number by the total number of possible poker hands. (The common pitfall in solving problems of this kind is to confuse permutations with combinations. Since the sequence of the cards is unimportant, we must deal only with combinations.) Let's begin by determining the number of ways (combinations) in which a pair of equal rank can be dealt. Take 7's, for example. There are four 7's in a poker deck. We are interested in finding the number of different pairs that can be made out of four 7's. This is equal to the number of combinations of 4 objects taken 2 at a time, or

$$_4C_2 = \frac{4!}{(4-2)! \times (2!)}$$

$$= \frac{4 \times 3 \times 2 \times 1}{2 \times 1 \times 2 \times 1}$$

$$= \frac{24}{4} = 6$$

So there are 6 ways in which a pair of 7's can be obtained. Since

43

the same reasoning holds true for all 13 ranks, there are $13 \times 6 = 78$ combinations in a poker deck of 2 cards having the same rank. Using the same principles, there are

$$_4C_3 = \frac{4!}{(4-3)! \times (3!)} = 4$$

ways of getting 3 queens—or 3 of any other rank. Since one of the 13 ranks was used in getting a pair, there are only 12 left, so the number of ways of getting a triple is equal to $12 \times 4 = 48$. Since each triple can be used with each pair, there are $78 \times 48 = 3,744$ ways of getting a full house.

The number of different poker hands is equal to the number of combinations of 52 cards taken 5 at a time. Using the formula for combinations,

$$_{52}C_5 = \frac{52!}{(52-5)! \times (5!)}$$
$$= \frac{52!}{47! \times 5!}$$
$$= \frac{52 \times 51 \times 50 \times 49 \times 48}{5 \times 4 \times 3 \times 2 \times 1} \quad \text{(canceling out the factors of 47!)}$$

The probability of being dealt a full house is equal to the number of favorable outcomes ($13 \times 6 \times 12 \times 4 = 3,744$) divided by the total number of possible outcomes. The probability P is, then,

$$P = \frac{13 \times 6 \times 12 \times 4 \times 5 \times 4 \times 3 \times 2 \times 1}{52 \times 51 \times 50 \times 49 \times 48}$$

After all possible cancellations are performed, this reduces to

$$P = \frac{6}{17 \times 5 \times 49} = \frac{6}{4165}$$

or about 1 chance in 700.

In a recent game of duplicate bridge, the author had the pleasant experience of being dealt 11 spades in one hand! What do you imagine the chances are of getting an 11-card suit? One in 1,000? One in 50,000? One in 100,000? All these guesses are far too low!

44

The number of combinations of 13 cards (all in 1 suit) taken 11 at a time is equal to $_{13}C_{11}$, or just 78 combinations. Since there are 4 suits, there are $4 \times 78 = 312$ ways of getting an 11-card suit. In contrast to this relatively small number, the number of possible bridge hands amounts to $_{52}C_{13}$ or precisely 635,013,559,600 different combinations! The chances of getting an 11-card suit, then, are a microscopic 1 in 2,000,000,000! Needless to say, I don't expect to be dealt a hand like that too often.

How did the ancients fix the direction of true north?

As man emerged from the Neolithic age, he began to set up crude monuments with which to fix the direction of seemingly important occurrences in the heavens. His reasons were logical and simple. He knew that there existed a regular succession of seasons and that these seasons were somehow connected with the changes he noticed in the location of stars and of the sun. If he could determine accurately the direction of such celestial phenomena as the setting of the sun, or the rising of a certain star, he could deduce the length of the year and the beginning and end of each of the seasons. When the ancient cities along the Nile were still young, man had determined the number of days in the year by fixing the rising of Sirius. In much the same way, ancient monuments like Stonehenge, and the remains of the Maya culture, show the direction of sunrise or sunset on the solstices by alignment with two pillars of unequal height. These directions may have been determined independently or by making a line at right angles to the north-south meridian.

Records passed down to us from ancient times illustrate the way in which the north-south direction was determined accurately. It was well known that the noon shadow was the shortest of the day and that it pointed directly toward the place in the heavens around which the stars revolve at night. To fix this direction accurately, it was only necessary to fix the direction of the midday shadow (Figure 10). This was done in the following way. A pole was placed in the ground, and a circle of appropriate length traced with a piece of cord around the pole. Since the shadow made by

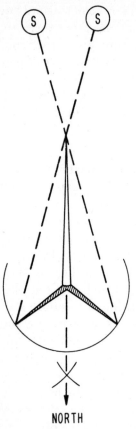

FIGURE 10. An ancient method of fixing the direction of the midday shadow.

the pole was variable in length, its tip would touch the circle at just two points during the day. Each of these points on the circle was marked, and an angle was formed by the two points and the shadow pole. The north-south direction was then determined by bisecting this angle. This was probably done at first by stretching a cord between the two points on the circle and then halving its length. At a later date, the more modern method of tracing arcs of equal length about the two points was probably used.

The east-west direction could have been fixed by tracing a line at right angles to the north-south meridian. Another method would have involved the annual change in direction of the setting or rising sun. The sun rises in the true east and sets in the true west on only two special days of the year—the vernal and autumnal equinoxes. On these days the night and day are of equal length. In the Northern Hemisphere the winter sun sets in the southwest, while the summer sun sets in the northwest. These extremes of the sun's directions are relatively easy to determine. The day of the spring equinox could have been determined as the day when the sun rises and sets midway between the winter and summer solstices. The east-west direction, of course, is the direction of the rising and setting sun on that day.

Why does a billiard ball often come to a complete stop after striking a second ball?

Anyone playing pool or billiards for the first time is struck by the tendency of the cue ball to come to a complete stop after striking a second ball. For this abrupt halt to take place, the motion of the moving ball must be along the line of centers of the two balls, and it must not have too much "english" or rotational velocity.

You can illustrate the principle involved with the help of several identical coins and a flat surface. Merely place several of the coins in a straight line (flat side down) with their sides touching. Then slide another similar coin into either end of the group, making sure that its motion is along the line of centers of the coins. The sliding coin will come to a complete stop, and the coin on the opposite end will move away from the group. All the intermediate coins will seem to remain stationary.

Paradoxes of this kind can be explained in part through the use of the physical concept of *momentum*.

Momentum is defined as the product of an object's mass and its velocity.

$$\text{Momentum} = m \times v$$

where m is the mass and v is the velocity of the moving object. The law of *conservation of momentum* states that the sum of the momenta of two bodies before collision is the same as the sum of their momenta after collision. If

47

v_1 = velocity of moving body before impact
v_2 = velocity of second moving body before impact
u_1 = velocity of first body after impact
u_2 = velocity of second body after impact
m_1 = mass of first body
m_2 = mass of second body

the law of conservation of momentum can be written

$$m_1 v_1 + m_2 v_2 = m_1 u_1 + m_2 u_2$$

In applying this formula to the billiard-ball problem, all the masses are the same ($m_1 = m_2 = m$) and $v_2 = 0$ since the second ball is stationary before impact. The formula then reduces to

$$m v_1 = m u_1 + m u_2$$

Since the mass m is common to all terms, we can divide the equation by m to give

$$v_1 = u_1 + u_2 \qquad (1)$$

Equation (1) tells us that the sum of the velocities after impact is equal to the velocity of the first ball before impact.

In his study of motion, Sir Isaac Newton observed that the *relative* velocity of two bodies after impact is proportional to the relative velocity before impact, but of the opposite sense. It has been found experimentally (for an elastic collision) that

$$v_1 - v_2 = -(u_1 - u_2)$$

The quantity ($v_1 - v_2$) represents the difference in velocity of two bodies before impact and, therefore, their relative velocity at that time. Similarly, $u_1 - u_2$ is their relative velocity after impact. The negative sign in the equation indicates that the *direction* of the relative velocity reverses after the collision. In other words, if v_1 is greater than v_2 (as it must be for a collision to occur), then u_2 will be greater than u_1—the second ball will move faster than the first after impact. In the billiard-ball problem, $v_2 = 0$ and the equation becomes

$$v_1 = -(u_1 - u_2)$$
$$v_1 = u_2 - u_1 \qquad (2)$$

Adding equations (1) and (2)

$$2v_1 = 2u_2$$
$$v_1 = u_2$$

Substituting this value for v_1 in equation (1) gives

$$u_2 = u_1 + u_2$$
$$u_2 - u_2 = u_1$$
$$0 = u_1$$

Since $u_1 = 0$, the first ball must come to rest after impact.

It might be argued that this is no proof at all since the equations used in the analysis were empirical in nature. We began by asking a question concerning the motion of billiard balls, and answered the question by introducing an empirical equation which states (mathematically) that billiard balls act in a certain way. The mathematical manipulations tell us *how* the experiment will come out but not *why* it comes out that way.

Such arguments, of course, are completely sound. Intuition and experience confirm Newton's principle concerning the reversal of direction of the relative velocity but, in the past, intuition and experience have been notoriously poor criteria upon which to base our ideas of nature. When we apply mathematics to the solution of a problem of the real world, we always make certain assumptions about that world—we construct a physical model. Newton's principle is a part of that model for the phenomenon we have considered. The mathematical solution is true, then, only to the extent that our physical model approximates the real state of affairs. In this sense, no fundamental law of nature can ever be proved mathematically. All such laws are empirical in nature and subject to revision when and if new contradictory observations are made. And can we ever be sure that we have included all of the significant factors in a physical theory?

What is the sum of the first *n* integers?

The sum of the first *n* integers can be represented by *S* as follows:

$$S = 1 + 2 + 3 + \cdots + (n - 2) + (n - 1) + n$$

where n is any integer. Now let's rewrite the equation for S *twice*, the second equation having the terms on the right-hand side in reverse order.

$$S = 1 + 2 + 3 + \cdots + (n - 2) + (n - 1) + n$$
$$S = n + (n - 1) + (n - 2) + \cdots + 3 + 2 + 1$$

If we now add these equations together, we obtain

$$2S = (n + 1) + (n - 1 + 2) + (n - 2 + 3) + \cdots + (n - 2 + 3) + (n - 1 + 2) + (n + 1)$$

which can be simplified to give

$$2S = (n + 1) + (n + 1) + (n + 1) + \cdots + (n + 1) + (n + 1) + (n + 1)$$

It's easy to see that there will be n of the binomials $(n + 1)$ on the right side of the equation, so

$$2S = n(n + 1)$$
$$S = \frac{n(n + 1)}{2}$$

If one wants to find the sum of the first 78 integers, he merely substitutes 78 and $78 + 1 = 79$ in the formula.

$$S = \frac{(78)(79)}{2} = 3{,}081$$

The sum of the first 78 integers is found to be 3,081 in less time than it would take to find a piece of paper large enough to do the sum the hard way!

What is compound interest?

Interest is a rental paid for the use of money. The *interest rate* is the amount of interest involved per dollar per unit of time. The most common unit of time is a year, but interest payments are often made on equally spaced dates throughout the year. The interval between interest payment dates is called the *interest period*. If *simple interest* is involved, the loan extends over one interest pe-

50

riod. For example, if you borrow $1,000 at 6 per cent interest for one year at simple interest, you would repay $1,060 ($1,000 + $60 interest) at the end of one year. If the transaction extends over several pay periods, the system of *compound interest* is usually used. Most savings banks use this system. Suppose a bank pays 4 per cent interest on deposits paid quarterly. This means that 1 per cent is paid each three months. If you were to deposit $100 at the start of an interest period, your account would be credited with $1 at the end of three months. Your bank balance would then read $101. At the end of the next quarter, an additional 1 per cent interest would be added and the balance would be $102.01 ($101 + $1.01). At the end of each period, the interest earned is added to the previous balance and the sum becomes the new principal for the next interest period. The compound interest for the two periods mentioned above is $2.01, while the equivalent simple interest for the same period of time would have been $2. The difference of 1 cent is small, but if we were to carry the process out over many pay periods, the difference would become considerable. To illustrate, suppose two banks advertise a 4 per cent interest rate. The first pays simple interest, while the second pays interest compounded quarterly. Which bank offers the greatest return on your savings? The second, of course. On a deposit of $10,000 for a period of 10 years, the second bank would actually pay you an additional $86.20 over that period of time.

It's possible to develop a formula which will enable you to calculate the size of your savings account after any length of time, at any desired interest rate. If P is the amount at the start of the first interest period, and r (expressed as a decimal) is the interest rate *per period,* the interest for the period will be rP. In compound interest this is added to the original amount to get the final amount

$$P + rP = P(1 + r)$$

The effect of the first interest payment is to multiply the principal by $(1 + r)$. During the next interest period the same thing will happen except that the new principal is $P(1 + r)$. The amount at the end of the second period is then

51

$$[P(1 + r)](1 + r) = P(1 + r)^2$$

The effect of each succeeding interest period is to put in another factor, $(1 + r)$. For n periods the formula becomes

$$\text{Final amount} = P(1 + r)^n$$

As n gets very large, it's probably easiest to use logarithms or an adding machine to determine the final amount. But even this becomes too burdensome for practical purposes so the results given by this formula have been tabulated in actuarial tables for various interest rates and many values of n. Such a table is called a *compound interest table.*

How many colors are there?

Most of us have seen a beam of light passed through a prism and separated into a rainbow of dazzling colors. It begins with red and ends with violet, and scientists call it the *spectrum.* There are obviously a great number of colors in the spectrum, but how many? If you have ever tried to mix a batch of paint to match an existing color, you know how tedious such an operation can be. With the help of a simple but fantastically precise mechanism, scientists have been able to break the visible spectrum down into 100,000 colors! This mechanism, known as the *diffraction grating,* was perfected in 1880 by Dr. Henry A. Rowland, Professor of Physics at Johns Hopkins University in Baltimore. As a matter of fact, today, some eighty years later, most of the world's supply of diffraction gratings is still manufactured in the basement of a building on the campus of this university.

The diffraction grating is nothing more than a thick and slightly concave mirror about six inches long. Although your eye would never notice it, a diamond needle has cut 30,000 equidistant lines to the inch on the face of the mirror. When a beam of light strikes the mirror, these 180,000 lines break up (diffract) the light into 100,000 distinguishable colors! Imagine the precision needed to produce 30,000 equidistant lines to the inch. This is equivalent to drawing about 100 perfectly straight equidistant lines on the edge of a page of this book! The machine that performs this operation is said to be the most precise mechanism in existence today.

How long does it take the moon to pass over a star?

It takes the moon, on the average, slightly more than $29\frac{1}{2}$ days to proceed through all its phases from new moon to new again. This period, called the *synodic month*, varies in length by more than half a day over many cycles. Since the average synodic month is usually less than a calendar month, the dates for the different phases are generally earlier in successive months.

To solve the problem at hand, however, we must use the *sidereal month*—the month of the stars. It takes just one sidereal month, or $27\frac{1}{3}$ days for the moon to make one revolution around the earth. At the end of this period of time, the moon has returned to nearly the same place among the stars. The moon has two apparent motions when viewed from the earth: it rises and sets daily circling westward around us; and it moves eastward much more slowly against the turning background of the stars. The latter motion is a result of the moon's revolution around the earth.

In the course of a day, the moon moves eastward among the stars $\dfrac{360°}{27\frac{1}{3}}$, or just about 13.2°. This amounts to an angular motion of $\dfrac{13.2°}{24}$ or 0.55° in 1 hour. If we calculate the moon's apparent diameter in degrees, it will be possible to determine the number of moon diameters that the moon moves in an hour. The diameter of the moon is 2,160 miles and its mean distance from the earth is 239,000 miles. Its apparent diameter in degrees is, then,

$$\frac{\text{Diameter of moon}}{\text{Circumference of moon's orbit}} \times 360°$$

$$\text{or} \frac{2,160}{2\pi \times 239,000} \times 360° = 0.51°$$

So the moon moves toward the east slightly more than its own diameter in 1 hour. This motion can be quite striking when the

53

moon passes over or *occults* a bright star. About an hour (or more precisely, 55 minutes) after the eastern edge of the moon has swallowed up the star, it reappears suddenly at the western edge. If the occultation is not central, however, the time of the eclipse will be correspondingly shorter.

How can you decide when to drop out of a poker hand?

The mathematical concept of *expectation* is extremely useful in determining when to continue or when to abandon an uncompleted gambling operation. Simply speaking, an expectation is the cash value of an uncompleted wager. Suppose that during the course of a wager, such as draw poker or a turkey raffle, the bettor assesses the probability of winning as 4 chances in 10 or 0.4. If he expects to win $10 in the event of a favorable outcome, his expectation is $0.4 \times \$10 = \$4$. In mathematical terms, the expectation is the product of the probability of winning times the possible winnings.

$$\text{Expectation} = \text{probability} \times \text{winnings}$$

For example, suppose a poker player plans to call a bet of $2 which would increase the pot (potential winnings) to $14. If he can see from the exposed cards that his probability of winning is 0.4, his expectation is $0.4 \times \$14 = \$5.60$. To get this expectation, he must pay $2. In other words, an expectation worth $5.60 can be purchased for $2; the expectation is worth more than the purchase price so he should call the bet.

To take another example, suppose your club is conducting a lottery in which the prize is worth $100. And suppose you hold 1 ticket out of a total of 50 issued. Your probability of winning is $\frac{1}{50} = 0.02$. Your expectation, then, is $0.02 \times \$100 = \$2$, which is the proper valuation of your ticket. If you decide to sell your ticket before the drawing, $2 is its fair market value.

What is applied mathematics?

Pure mathematics is concerned with the deductions that can be made from a few defined concepts and a number of agreed-upon axioms. Starting with the point, line, and plane, and a number of simple axioms, the great body of knowledge known as geometry has been developed. Applied mathematics, on the other hand, involves giving *physical significance* to the concepts of pure mathematics so that the mathematical theorems developed will be useful in scientific work. Take π, for example. In pure mathematics, π is the ratio of the circumference of any circle to its diameter. In applied mathematics $\pi = 3.1416$, which is the ratio of the circumference of any circle to its diameter, correct to four decimal places.

Applied mathematics in the modern sense is based on an ingenious marriage of geometry and algebra created independently by two great French mathematicians of the seventeenth century, René Descartes and Pierre de Fermat. This new branch of mathematics, known as *coordinate geometry*, created a revolutionary and effective way of representing and analyzing curves. Its invention is probably one of the most significant events in all scientific history.

During the early part of the seventeenth century, mathematics was still essentially the geometry of Euclid garnished with a bit of algebra. But Euclidean geometry was confined to figures that could be drawn with straight lines and circles—figures that were hardly adequate to deal with the parabolas, ellipses, and hyperbolas of the new technologies. In 1609, Kepler had shown that the planets move in elliptical paths instead of the circles of Hipparchus. Parabolas were important because they closely describe the paths of projectiles such as cannon balls. The curvature of lenses was important in the study of telescopes, microscopes, and of the human eye itself. At every turn, scientists were hampered by a body of mathematical knowledge that was incapable of solving their most difficult problems. At this point, Descartes and Fermat stepped into the picture.

To understand their accomplishment best, let's follow the reasoning of Descartes. Suppose we start with any curve such as the curve *APB*, shown in Figure 11. This curve can be thought of as being generated by a point *P* on a vertical line *PQ*. As the line moves to the right, *P* moves up or down in accordance with some prearranged plan to trace out the shape of the curve. Descartes used algebra to define this prearranged plan.

As the line moves to the right, its distance from a point such as

55

O can be used to describe its position. This distance, OQ, can be denoted by x. The location of the point P on the vertical line can be specified by giving its distance above the horizontal line OQ. This distance can be denoted by y. So for every position of P, there is a corresponding number for x and a corresponding number for y.

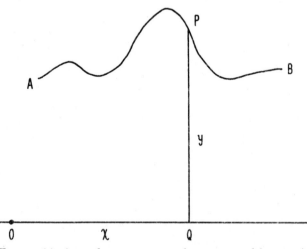

FIGURE 11. Any plane curve can be generated by a point P, which moves about in accordance with some prearranged plan.

We can now go a step further by systematizing our thoughts. Our coordinate geometry begins with a straight horizontal line which we can call the X-axis, since the values of x are measured along this line. We also have a point O, the *origin*, from which our measurements are taken (Figure 12). Let's draw a vertical line through O which we can call the Y-axis. Now if P is any point on the curve, then there are two numbers that completely describe its position. The first is x, the distance from the origin to the point directly beneath P, and y, the height of P above the X-axis. The number x is measured along the X-axis and y is measured along the Y-axis. These numbers are usually called the *coordinates* of P and are generally written (x,y). The system of axes described

56

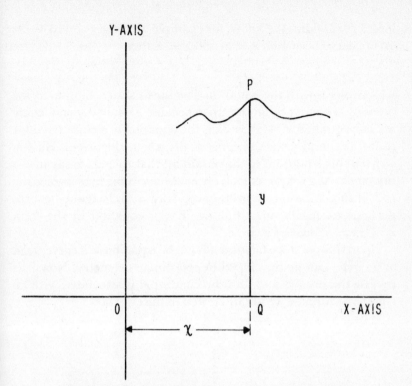

FIGURE 12. A curve located on the coordinate axes.

above is called the *rectangular coordinate system*. If P is to the right of the Y-axis, its x-value is taken to be positive; and if it is to the left of the Y-axis, its x-value is negative. Similarly, the y-value is positive if the point is above the X-axis; and negative, if it is below.

Let's see how this system can be used to develop the equation of a simple curve—the circle. Suppose the circle has a radius of 5. We can draw such a circle on the rectangular coordinate system so that it is symmetrical about the origin, O. Let P be *any point at all* on the circumference of this circle and let x and y be its coordinates, as shown in Figure 13. From the Pythagorean theorem, we know that

$$x^2 + y^2 = 25$$

This equation holds for every point on the circle, since P was se-

57

lected at random. We know, for example, that the point (3,4) lies on the curve because if $x = 3$, and $y = 4$, then

$$3^2 + 4^2 = 5^2$$

which agrees with the equation developed above. Similarly, the point (2,3) cannot be on the circle because $2^2 + 3^2$ does not equal 5^2. The equation $x^2 + y^2 = 25$ is the equation of a circle having a radius of 5 and drawn symmetrically about the origin. Having written this equation, we have said all that there is to say about this circle. As a matter of fact, once we have come to recognize the *form* of this equation we need never draw a circle again—except perhaps to clarify our thinking. Every equation in the form $x^2 + y^2 = R^2$ *must* be a circle.

Up to this point we have seen how the equation of a curve, such as a circle, can be developed in coordinate geometry. Now let's reverse the process and find the curve that is associated with an equation. Suppose we are interested in the equation

$$y = x^2$$

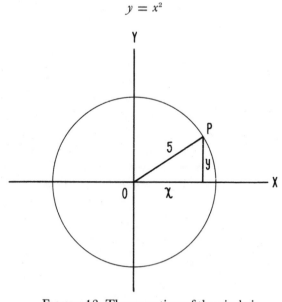

FIGURE 13. The equation of the circle is $x^2 + y^2 = 5^2$.

58

What curve is associated with this equation? To find out, let's assume successive numbers for x and determine the corresponding numbers for y that *satisfy* the equation. If $x = 1$, for example, $x^2 = 1$ and $y = 1^2 = 1$. If $x = 2$, then $y = 2^2 = 4$. Note also that x can be negative as well as positive since x^2 will yield the same result in either case. Tabulating a few corresponding numbers for x and y we get:

x	y
0	0
± 1	1
± 2	4
± 3	9
± 4	16

Each pair of equivalent numbers for x and y corresponds to a point P on the curve that we are looking for. Take the point $(1,1)$ in Figure 14. The point for these coordinates is found by counting 1 unit to the right of the origin (since x is positive), and 1 unit above the X-axis (since y is positive). Similarly, $(-3,9)$ is located 3 units to the left of the origin and 9 units above the X-axis. The complete curve is drawn in Figure 14, although it is understood that it extends indefinitely in the upward direction both to the left and to the right. This curve is symmetrical about the Y-axis and it can be proved to be a parabola.

Before we leave the subject, it might be worth while to illustrate one practical application of coordinate geometry. Suppose it's desired to find the points of intersection of two curves:

$$y = x^2 \quad \text{(a parabola)}$$
$$y = 2x \quad \text{(a straight line)}$$

We could, of course, draw these curves precisely to scale and read the points of intersection from the drawing. But this is tedious work and we would never know the degree of accuracy of our results. We can find the answer more simply by subtracting the second equation from the first and solving the new equation that results.

59

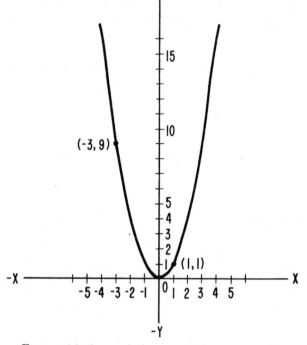

FIGURE 14. A parabola, $y = x^2$, in the coordinate geometry of Descartes.

$$y = x^2$$
$$-y = -2x$$
$$y - y = x^2 - 2x$$
$$0 = x^2 - 2x$$
$$0 = x(x - 2)$$
$$x = 0$$

or $x = 2$

The two solutions for x tell us that there are two points of intersection of the equations in question: the first is at the point which has an x-value of 0 and the second is at the point where $x = 2$. From the original equations, we find the corresponding y-values to be 0 and 4. The points of intersection are, therefore, (0,0) and (2,4).

60

As one goes deeper into the new geometry of Descartes and Fermat, he finds that each curve has its own equation; an equation that uniquely describes all the points that lie on that curve. Conversely, each equation involving x and y has its own peculiar curve which can satisfy no other equation. This association of curve and equation was a new concept that combined the best of Euclid with the best of algebra. It provided a new and extremely valuable mathematical tool for use in studying many kinds of geometrical and scientific problems.

What are cardinal and ordinal numbers?

Two principles permeate all mathematics and, for that matter, all areas of exact thought—*correspondence* and *succession*. It's not too surprising, then, to find these elements woven deeply into the very fabric of our number system. A hundred times a day we engage in measuring and counting, little realizing that we are performing two distinct kinds of operations. The distinction between correspondence and succession is well obscured by a marriage of prehistoric age.

When we use a cardinal number such as 6, or 8, or 3, we make use of a most rudimentary faculty which Dantzig has called *number sense*. This ability enables primitive man to recognize the difference between two piles that contain different numbers of objects. The important point here is that this recognition is accomplished without the ability to count. The principle involved is that of correspondence. There is a similarity between a pair of wings and a nest containing two eggs; or between the legs of a horse and a group of four trees. By establishing a correspondence between the objects in a pile and some *familiar model* it is possible to use the concept of number without resorting to the artifice of counting. When we enter a classroom, for example, we notice two collections: the chairs and the students. *Without counting* either the students or the chairs we can determine whether the two collections are equal, or whether one is greater than the other. This ability to discern relative quantity is accomplished through the use of a process of placing objects

61

in *one-to-one correspondence.* What we have done is to assign to each member of one collection, the chairs, a member of another collection, the students. This shows clearly whether an equality exists, or whether one collection is greater than the other. This concept of matching or tallying is called *cardinal* number. Early man kept track of his herds and other possessions by means of notches cut in a stick, or pebbles gathered in a pile. Hence, tally and calculate are derived from the Latin, *talea,* cutting, and *calculus,* pebble.

The next step in comparing collections is the establishment of a group of standard, or *model* collections, each of which typifies a particular cardinal number. In this way, "legs" might stand for two, "clover leaf" for three, "horse" for four, "hand" for five, etc. Determining the cardinal number for an unknown quantity might then be reduced to finding a model collection which can be matched equally with the given unknown collection. In time, a number word, "hand" for example, would come to be synonymous with the corresponding cardinal number, even though the passage of time had erased its original meaning.

In spite of the importance of the cardinal numbers, they represent at best a rudimentary form of number sense. If man had been limited to their use, his intellectual progress would have been modest at best. The extraordinary progress of the human race depended upon the artifice of counting, which, in turn, depends upon the principle of succession.

In order to create the concept of counting, we must have more than a large group of models, or number symbols. Counting requires a number system, in which each number symbol is arranged in an ordered sequence. One, two, three, four, five, ... This sequence implies that each symbol is precisely 1 unit greater in magnitude than its predecessor; we have produced an *ordered succession* of numbers—the *ordinal* numbers. Having created this system, *counting* the members of an unknown collection merely involves assigning to each member an ordinal number in ordered succession until the unknown collection is exhausted. The ordinal number of the collection is then the last ordinal number to be used.

In practice we are usually interested in the cardinal number, but we determine this number by *counting*, not by searching for a model with which to match it. We have learned to pass from cardinal to ordinal numbers with such facility that we take this understanding for granted. But cardinal numbers alone can never produce an arithmetic since calculation involves the tacit assumption that we can always move from one number to its successor—the ordinal concept.

Is it possible to multiply on one's fingers?

Although finger counting is now a forgotten craft among most civilized peoples, it played a major role in the art of calculation as recently as the sixteenth century. No early textbook on arithmetic would have been considered complete without a full explanation of the techniques involved. In time, the advent of cheap writing materials and the simplified arabic number system rendered the art obsolete, so we tend to minimize its importance in the development of number. Yet, in the form of his 10 fingers, man found the built-in abacus that enabled him to move effortlessly and almost imperceptibly from primitive to advanced forms of mathematical thought. These 10 fingers enabled him to discover the connection between quantity and the counting process.

If a man wants to describe a collection of 3 items, he can hold up 3 fingers—the model of the cardinal number 3. If he wants to count the same collection, he turns up the same 3 fingers, but this time *in succession*—in the ordinal system.

Once the numbers are connected in an ordered sequence, it's possible to develop many ingenious rules for adding and multiplying on one's fingers. Suppose, for example, that you had learned the multiplication table only through 5; how would you multiply 8 × 7? Here's the rule: Subtract 5 from each number to be multiplied, giving 3 and 2 in this example; bend down 3 fingers on the left hand and 2 on the right hand; the sum of the bent-down fingers (3 + 2 = 5) gives the tens figure of the answer, and the product of the unbent fingers (2 × 3 = 6) gives the units figure of the answer. The reverse of this procedure can be used in division.

Historians tell us that man must have developed his number concept and its language many thousands of years before he had learned to write. Except for the number *five*, the original meaning of number words in the civilized languages has been lost in antiquity. In this one instance, the Sanskrit *pantcha*, five, bears a remarkable similarity to the Persian *pentcha*, hand. While the number words have shown great stability over long periods of time, other aspects of a language have always been subject to radical change. This undoubtedly accounts for our inability to trace the original meanings of the number words.

What are the chances of being invited to 2 birthday parties on the same day?

Have you ever been invited to 2 birthday parties on the same day? If not, have you ever wondered what the chances are of receiving such double invitations? If you have only 24 friends that are close enough to invite you to their parties, you might assume that the chances of 2 coincident birthdays are highly remote. There are 365 days in the year on which their birthdays may fall and there would seem to be little chance that 2 of the 24 would fall on the same date.

Unbelievable as it may seem, however, intuition is completely wrong in this situation. The truth of the matter is that the chances are *slightly better than even* that 2 or more of your 24 friends will have birthdays on the same date!

If you're still in doubt, the statement can be verified through the use of the rules of probability. As with many such problems, the method of attack is to turn the problem around and ask the probability that each of the 24 persons has a different birthday. Here's how it's done. The first person in the group can have any birthday at all. Now, what are the chances that the second person has a different birthday? Since there are 365 days in the year and 1 of them has been used by the first person, the chances are 364 out of 365—the probability is $\frac{364}{365}$ that the birthdays are different. Similarly, the probability that the third person has a different birthday is $\frac{363}{365}$ since 2 days of the year have been excluded. The probabilities that each of the remaining persons will have a different birthday are $\frac{362}{365}$, $\frac{361}{365}$, $\frac{360}{365}$, $\frac{359}{365}$ and so on down to the twenty-fourth person for whom the probability is

$$\frac{365 - 23}{365} = \frac{342}{365}$$

The probability that no 2 of the persons have coincident birthdays is the product of the separate probabilities

$$\frac{364}{365} \times \frac{363}{365} \times \frac{362}{365} \times \cdots \times \frac{342}{365} = 0.46$$

The product is equal to 0.46 which indicates that there are 46 chances in 100 that no 2 of your 24 friends will have birthdays on the same date. On the other hand, there are $100 - 46$, or 54 chances in 100, that 2 or more will.

The problem of coincident birthdays is one more example of how completely wrong our intuitive judgments can be.

Which rectangle is the most beautiful?

Emerson said that beauty is, among other things, "the mean of many extremes." In telling about the building of the great pyramid of Gizeh, about 4700 B.C., the papyrus of Ahmes reads, "The sacred quotient, seqt, was used in the proportions of our pyramids." The Greeks called this sacred quotient the "golden section," and in medieval times it was known as the law of divine proportion. But whatever its name, this number can be obtained by dividing a line of any length into two parts such that the lesser is to the greater as the greater is to the whole length. The ratio of these parts will be $\frac{1}{2}(\sqrt{5} - 1) = 0.618$, or $\frac{1}{2}(\sqrt{5} + 1) = 1.618$. These ratios can be derived easily with the help of algebra. Divide a line into two parts and set one part equal to x and the other equal to $1 - x$. Then

$$\frac{1-x}{x} = \frac{x}{1} \qquad \text{or} \qquad x^2 + x - 1 = 0$$

Completing the square,

$$x^2 + x + \frac{1}{4} = 1 + \frac{1}{4}$$

$$\left(x + \frac{1}{2}\right)^2 = \frac{5}{4}$$

$$x + \frac{1}{2} = \pm \frac{\sqrt{5}}{2}$$

$$x = \frac{\sqrt{5} - 1}{2} = 0.618 \cdots$$

$$x = \frac{-\sqrt{5} - 1}{2} = -1.618$$

Since the minus sign in the second root has no aesthetic significance we may neglect it, giving the two ratios mentioned earlier. Architects and artists since ancient times have used the "golden rectangle" with a ratio of base to height of 1.618 in proportioning rooms, buildings, temples, picture windows, and hundreds of other items.

Why does the moon play the major role in tide formation?

At first glance it would seem that the moon's great influence on the tides is due to its proximity. The moon is only 240,000 miles away, whereas the sun is 93,000,000 miles away. It's only reasonable, then, to expect the moon to have the greater effect (Figure 15). Distance, however, is only part of the story. If the sun were about 20,000,000 miles from the earth, the roles of the sun and moon would be reversed and the sun would be twice as effective in tide formation. High tides would always occur at noon and midnight and low tides would come at 6 P.M. and 12

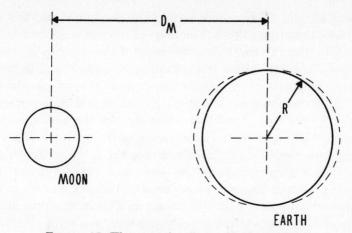

FIGURE 15. The moon's role in tide formation.

P.M. To understand why this is so, we must keep in mind the fact that tides occur simultaneously on opposite sides of the earth. It's easy to understand why water builds up on the side of the earth under the moon, but the bulge on the opposite side of the earth is less obvious. The moon's attractive force at the center of the earth is greater than at the distant surface, and less than at the near surface. The water on the near surface is attracted toward the moon, while the water on the other side of the earth is, in effect, released. As the earth rotates on its axis, the two mounds of water maintain their relationship to the moon to form two tidal waves that seem to move around the world about once a day.

The magnitude of each of these bulges depends upon the *difference* between the attractive forces at the center of the earth and at its surface. If there were no difference in these forces, all parts of the earth would be attracted moonward by the same gravitational force and there would be no tides.

The force of gravitation between two bodies is given by the expression

$$F = \frac{M_1 \times M_2}{D^2}$$

where M_1 and M_2 are the masses of the two bodies and D is the distance separating them. The mass of the sun is about 300,000 times that of the earth and the mass of the moon is $\frac{1}{100}$ times that of the earth. Using this information, and the radius of the earth (4,000 miles), it can be shown that the *difference* in the moon's attractive force at the earth's center and at its surface is about 3.3 per cent. The similar difference for the sun's attractive force is 0.0086 per cent. The *magnitude* of the sun's gravitational force, however, is about 200 times as great as that of the moon. The much greater mass of the sun can't quite overcome the moon's shorter distance, however, and the moon ends up the winner by almost 2:1. In the moon's case there is a force difference of 3.3 per cent. In the sun's case there is a force difference of 0.0086 per cent of something that is 200 times as great. But $0.0086 \times 200 = 0.0172$, or 1.72 per cent, which is just about half as great as 3.3 per cent.

Why is it wrong to divide by zero?

Check over the operations of this old favorite and see if you can find the fallacy:

Let:	$a = b$
Multiply both sides by a:	$a^2 = ab$
Subtract b^2 from both sides:	$a^2 - b^2 = ab - b^2$
Factor:	$(a + b)(a - b) = b(a - b)$
Divide both sides by $(a - b)$:	$(a + b) = b$
But $a = b$; so:	$2b = b$
Divide both sides by b:	$2 = 1$

The trouble comes about, of course, when we divide both sides by $(a - b)$, which is equal to zero. While the fourth line is correct, the fifth is not. Division by zero is frowned upon in better mathematical circles because every time it's tried, something absurd like $2 = 1$ results.

Somewhere back in the third or fourth grade you were probably admonished: "Thou shalt not divide by zero." If you were curious enough to wonder why, you were probably told that it just isn't done; division happens to be a process which does not include zero among the possible divisors. Luckily, we can now manage a somewhat better explanation than that.

Division is really the inverse of multiplication. When we say that $10 \div 2 = 5$, we also imply that $2 \times 5 = 10$. When we seek the answer to the problem $10 \div 2 = ?$, we ask ourselves to find a number which when multiplied by 2 equals 10. Similarly, when we write $10 \div 0$, we want a number that gives 10 when multiplied by 0. There is no such number. So division by zero yields a number that mathematicians have not, as yet, been able to find.

Looking at the situation another way, we are forced to admit that zero times any number is zero, or $0 \times a = 0$. Consequently, the symbol $\frac{0}{0}$ stands for *any* rational number. On the other hand, the equation $0 \times a = b$ is impossible if b is other than zero. The symbol $\frac{b}{0}$, then, can stand for *no* rational number. On the basis of such reasoning, we say that $\frac{0}{0}$ is a possible but ambiguous symbol, while $\frac{b}{0}$ is impossible since it represents no rational number.

How can arithmetical calculations be checked?

Unfortunately, most of us are not very good at arithmetic. As a matter of fact, even mathematicians sometimes joke about the difficulty they encounter in balancing their bank accounts. While that problem may transcend mere arithmetical proficiency, it is still useful to be able to check the accuracy of computations.

One method of doing this makes use of an ancient *rule of nine* that dates back to the days of the abacus. It depends on the fact that when 9 is divided into 10, 100, 1,000, or any power of 10, it always leaves a remainder of 1. This is the reason, by the way, that $\frac{1}{9} = 0.111111 \cdots$ with as many 1's as you care to write. For this reason, 9 divides a number (in the decimal system) only if it also divides the sum of the digits of that number. Take 4,275, for example. Nine divides 4,275 exactly 475 times. The sum of the digits of 4,275 is $4 + 2 + 7 + 5 = 18$, which is also divisible by 9. And carrying it one step further, $1 + 8 = 9$, which is also divisible by 9. The rule which makes use of this fact can best be illustrated by an example. Suppose we multiply 897 by 67 and obtain 60,099. To check our results, we add the digits of 897, $8 + 9 + 7 = 24$, and once again, $2 + 4 = 6$. Then we do the same for 67, which gives us 13, and finally, 4.

$$
\begin{array}{c}
897 \rightarrow 24 \rightarrow 6 \\
\underline{67 \rightarrow 13 \rightarrow 4} \\
\begin{array}{ll}
6279 & \quad 24 \qquad 6 \\
5382 &
\end{array} \\
\underline{} \\
60099 \rightarrow 24 \rightarrow 6
\end{array}
$$

Now multiply the 6 and 4 to get 24, which gives 6 for the sum of its digits. When we add the digits of the product, 60,099, we obtain 24, and finally, 6, which agrees with the 6 obtained above.

By successively adding the digits of a number until a single digit is obtained we are merely determining the *amount by which that number differs from a multiple of 9*. Take 24. Adding $2 + 4$ gives 6, which is the amount by which 24 differs from 18, which is a multiple of 9. The *rule of nine* can therefore be stated as follows: The product of any two numbers differs from a multiple of nine by the same amount that the product of the sums of the digits in the two numbers differs from a multiple of nine. In using this rule, however, you should keep in mind that two digits might have been transposed in the answer, or a zero added or lost. These are common mistakes that the check will not catch.

The rule of nine can also be used for addition and subtraction. If we are checking addition, we add the sums of the digits of the two numbers (6 and 4 in the example); if we are checking subtraction, we subtract them. When checking division, we reduce the dividend, divisor, quotient, and remainder to their equivalent single digit as above. Then we apply the standard

70

rule for checking long division: The dividend equals the divisor times the quotient plus the remainder.

This was the ancient method of checking work done on the abacus known as *casting out nines*. It depends, as we have said, on the fact that if a number is divisible by 9, the sum of its digits is also divisible by 9.

How is a flat map made to conform with the spherical earth?

To be of practical use, maps and charts must be printed on flat pieces of paper, whereas the earth's surface is very nearly a perfect sphere. So when large areas are to be mapped, the cartographer encounters severe complications. The conventional method of locating points on the earth's surface is by means of *latitude* and *longitude* measured in *parallels* and *meridians*. Taken together, these imaginary circles form a spherical *grid*. The basic problem, then, is to transfer this grid from the sphere to a flat surface. There are many ways to do this, and all are called *projections*.

Perhaps the earliest map known today was made in Babylon about the year 3800 B.C. For the next 5,000 years or so, scholars worked hard at map making with notable lack of success until a Flemish mathematician and geographer named Gerhard Kremer solved the problem once and for all. Since it was fashionable in those days for scholars to translate their names into Latin, his name became Gerhardus Mercator—and in that form became immortal. Mercator's projection is used on most of our maps and nearly all our charts. It is still the best and most widely used form of map making.

Mercator reasoned that you can't transfer a globe to a flat sheet of paper, *but you can transfer it fairly well to a cylinder.* While such a procedure would produce many errors, the errors would all be consistent and in accordance with a prearranged plan. Any navigator, knowing the plan, could read the map.

Here's how Mercator made his projection. Imagine yourself to be at the center of a transparent globe, the surface of which is marked with circles of latitude and longitude. Imagine further

71

that a sheet of paper is wrapped around the globe to form a cylinder touching the earth all along the equator. The cylinder of paper, of course, is open at both ends. As you look in various directions from the center of the globe, your eye "projects" the latitude-longitude grid from the globe to the paper. The equator becomes a circle on the paper at the line of tangency of globe and cylinder. Parallels of latitude are projected as circles parallel to the equator. Meridians of longitude are projected as straight lines perpendicular to the parallels. When the cylinder is cut along one of the meridians of longitude, it flattens out to reveal a pattern of meridians and parallels which are mutually perpendicular. The parallels of latitude are parallel horizontal lines and the meridians of longitude are parallel vertical lines.

The major objection to the Mercator projection is the distortion it generates in the size of near-polar regions. Greenland, for example, is portrayed larger than South America, although it is actually only one-eighth as large. But this kind of error is only important if one regards a map as a picture. Considered as a coded message, it provides accurate and easy-to-understand information needed to move about efficiently on our globe. And that, after all, is what a map is for.

What is an annular eclipse of the sun?

Like all opaque objects in sunlight, the earth and moon cast their shadows away from the sun. Each heavenly body is accompanied by its cone-shaped and ordinarily invisible shadow as it follows its course through the sky. But since the moon continually circles the earth, it's possible for one of these bodies to enter the shadow of the other. When the moon is darkened by the earth's shadow, we have a *lunar eclipse;* and when the earth is darkened by the moon's shadow, we have a *solar eclipse.* An eclipse can occur only when the earth and moon are in line with the sun, so a lunar eclipse will occur always during the phase of the full moon, while a solar eclipse requires the presence of the new moon.

The shadows of the earth and moon are represented diagram-

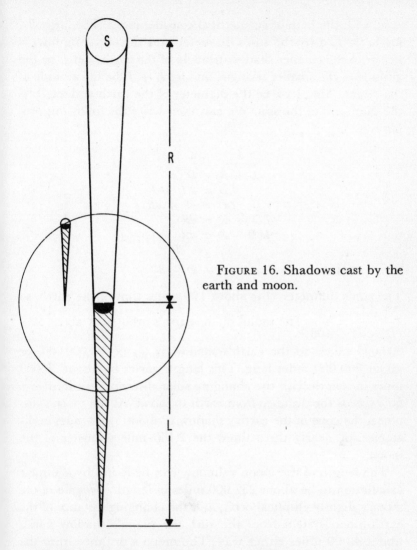

FIGURE 16. Shadows cast by the earth and moon.

matically in the accompanying sketch (Figure 16), but it isn't practicable to show the figures in correct proportion. The sun's diameter is about 109 times as great, and the distance of separation some 11,700 times as great as the diameter of the earth. The true length of the earth's shadow can be found easily, how-

73

ever, with the help of geometrical considerations. The large triangle formed by the sun's diameter and the two long lines is *similar* to the smaller shadow-triangle of the earth. Let L be the altitude of the smaller triangle, and let $L + R$ be the altitude of the larger. Also, let d be the diameter of the earth and let D be the diameter of the sun. We can then write the following proportion:

$$\frac{L}{L + R} = \frac{d}{D}$$
$$LD = d(L + R)$$
$$LD = dL + dR$$
$$LD - dL = dR$$
$$L(D - d) = dR$$
$$L = \frac{dR}{D - d}$$

The sun's diameter D is about 109 times that of the earth, so $\frac{d}{(D - d)} = \frac{1}{108}$. The radius of the earth's orbit R, is about 93,000,000 miles, so the earth's shadow is $\frac{1}{108} \times 93,000,000$, or about 860,000 miles long. This length varies by about 30,000 miles in length with the changing solar distance, but it always far exceeds the distance from earth to moon. At the moon's distance, the cone of the earth's shadow is about 5,700 miles in diameter, or nearly three times the 2,160 mile diameter of the moon.

The length of the moon's shadow can be found by a similar calculation to be about 232,000 miles in length. Because of the moon's slightly elliptical orbit, and the changing distance of the earth-moon system from the sun, the moon's shadow varies about 4,000 miles either way. The moon's distance from the earth ranges from 222,000 miles at *perigee* (its closest advance) to 253,000 at *apogee*. The *mean distance* of the moon is 239,000 miles, or somewhat greater than the average length of the moon's shadow. For this reason, its shadow usually falls short of the earth. Under these conditions, a person situated in line with the moon and sun will see a bright ring, or *annulus,* and the eclipse is said to be *annular.* When conditions produce a lunar shadow that is longer than the moon's distance, the diameter of the moon appears to be greater than that of the sun, and the eclipse is total.

The longest lunar eclipse, when the moon passes centrally through the earth's shadow, lasts about three hours and forty minutes. This period includes the partial phases when the moon is entering or leaving the shadow. The period of totality can be as great as one hour and forty minutes, preceded and followed by partial phases of about an hour's duration. Most lunar eclipses are not central, however, and are of shorter duration.

A lunar eclipse can be seen from any part of the nighttime side of the earth plus any part that is turned into view of the moon while the eclipse is going on.

A solar eclipse, on the contrary, can be seen from only a relatively small portion of the earth. As we have seen, the moon's shadow is usually not long enough to reach the earth. When it is long enough, the area covered by the shadow rarely exceeds 150 miles in diameter. In some solar eclipses, it becomes vanishingly small. Around this small area of the earth in which the eclipse appears total or annular, there is a much greater area in which a *partial* solar eclipse can be seen. This area may be 2,000 or 3,000 miles in diameter. Sometimes, only the partial phase can be seen. This happens when the axis of the moon's shadow is off to one side of the earth.

While a lunar eclipse may last for hours, a solar eclipse is quite a fleeting thing. As the moon revolves around the earth, its shadow sweeps in an easterly direction. While the earth also rotates toward the east, the relative speed of the shadow varies from 1,000 to 5,000 miles per hour, depending upon the latitude of the observer, and being greatest near the poles. Because of this great speed and the small size of the shadow-dot on the earth, a total solar eclipse can't last much more than seven and a half minutes at any particular place. The partial phase may last as long as

75

four hours or so, although it usually takes much less time.

A total solar eclipse is a truly striking occurrence. The partial phase begins with a black notch at the sun's western edge. As the moon occults more and more of the sun's disk, the sun becomes a crescent of light, resembling the moon going through its phases. Finally, only a thin crescent remains and the sky and land assume a pale, unfamiliar hue, for the light from the sun's rim is redder than our familiar sunlight. Even this faint and strange illumination fades rapidly as the total eclipse approaches. The air becomes chilled and birds and animals alike seem bewildered. Certain flowers begin to close for the "night." Then with surprising suddenness the last sliver of light disappears. With the sudden change to darkness, the sun's filmy corona bursts into view like a giant halo reaching out tenuously from behind the darkened moon. The filmy petals and streamers of the corona accent the planets and bright stars that usually come out during totality. This has made it possible for scientists to confirm that there are no planets of any consequence nearer to the sun than Mercury. In addition, the total eclipse of the sun has made possible one of the confirmations of Einstein's theory of relativity. Precise measurements show that the stars near the edge of the eclipsed sun are displaced outward, as they should be according to the theory.

If the moon's orbit coincided with the plane of the earth's orbit, called the plane of the *ecliptic,* there would be two eclipses each month. An eclipse of the sun would occur at the time of the new moon, and a lunar eclipse when the moon is full. But the plane of the moon's orbit is inclined 5° to the ecliptic plane, so the moon usually passes north or south of the line joining earth and sun. Even so, three eclipses of the moon are possible in the course of a year, though a whole year may pass without a single one. The sun has at least two and may have as many as five eclipses in a given year, although the greatest number of lunar and solar eclipses can't occur in the same year. The minimum number of eclipses in a year, then, is two, both of the sun. The maximum number is seven, either two of the moon and five of the sun, or three of the moon and four of the sun.

76

Who first measured the moon's distance from the earth?

We have seen that the ancient Greeks created in a few short centuries the great body of mathematical knowledge that we call Euclidean geometry. They created mathematics in the sense in which we use the term today. They insisted on deduction as the exclusive method of proof and much preferred abstract concepts to the particular. Then, about 325 B.C., Alexander the Great conquered all Greece, Egypt, and the Near East, and proceeded to unify his conquests into a vast and powerful empire. He founded the city of Alexandria and established it in place of Athens as the capital of the ancient world. Alexandria, located as it was at the junction of Europe, Asia, and Africa, became the commercial and cultural center of the entire ancient world. As a result, the city became a truly cosmopolitan center of learning as Alexandrian traders imported knowledge that had been acquired all over the world. Ptolemy the First, one of Alexander's able generals, took over control of Egypt on the death of the conqueror and established a large museum and library. It was here that the great minds of the time assembled to provide the cultural environment so necessary for the next important advance in mathematics.

While the Greeks were interested primarily in abstract forms of mathematics, the Alexandrians were interested in measuring things. As a result, they created and applied many forms of indirect measurement. Euclid knew that the areas of two similar triangles are in the ratio of the squares of corresponding sides, but the Alexandrians discovered how to determine the area of any one single triangle by multiplying the base times one-half the altitude. This is certainly much more useful, in a physical sense, than the alternative method of placing many small squares within the area to be measured, and then counting up the number of squares used.

This sort of indirect measurement was a great practical achievement of the Alexandrians. Eventually, they were able to measure by indirect means the diameters of the earth, sun, and moon, and the distances to the sun, moon, and planets. This is quite an accomplishment for a civilization removed by so short a length of time from almost complete mathematical ignorance.

77

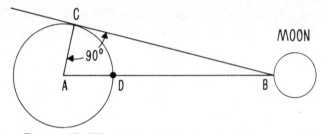

FIGURE 17. Hipparchus measures the distance to the
moon.

It was in the second century B.C. that Hipparchus, the great
Alexandrian mathematician and astronomer, calculated the
moon's distance as approximately a quarter of a million miles. His
estimate was only about 5 per cent off from the true distance. His
method is essentially the one used in finding the height of a cliff,
or building, by measuring the angle between the top and the base
from two positions a known distance apart. In practice, this becomes a bit complicated, but we can grasp the underlying principles by using a slight simplification that contains the essence of
Hipparchus' method. Let's suppose that the measurements are
made from two points on the equator, at a time when the moon
is directly over one of these points, i.e., at its zenith. This is point
D in Figure 17. At the same instant, another observer at point C
watches the moon rise on the horizon. A theorem of geometry tells
us that angle C is a right angle, so we may write

$$\text{Cosine } A = \frac{AC}{AB}$$

Since Hipparchus had already invented latitude and longitude, it
was a simple matter for him to determine angle A. A reasonable
value for A is $89°3'$. He had also made up the first cosine tables,
so he knew that cosine $89°3' = 0.01658$. In addition, he had previously measured the radius of the earth, AC, as about 3,950 miles.
Using these numbers in the equation, we find that

$$AB = \frac{3,950}{0.01658} = 238,000 \text{ miles}$$

That is, the distance from the moon to the center of the earth is just under a quarter of a million miles.

In general, we can measure the distance to any inaccessible point by starting with a readily measurable distance on the surface of the earth. Hipparchus used this technique to measure, successively, the height of a mountain, the radius of the earth, and the distance to the moon. Other Alexandrians determined the size of the moon and the approximate distance to the sun. These were the first daring voyages of man's mind into the great expanse of outer space. They represented an epochal advance in man's knowledge of the universe about him.

What are mathematical formulas?

Very little was accomplished in a scientific sense until men noticed that nature behaves in systematic and predictable ways. The ancients invented a method of logical thought and used it to deduce the secrets of nature. When the results of careful observation disagreed with their world of logic, then observation must be wrong. Modern science began when men like Galileo, Copernicus, and Kepler—at great personal risk—rejected this socially and theologically popular concept in favor of the truth. But whenever science looks for truth it seems to find order; and in order, mathematical relationships.

Galileo and his students used hourglasses to measure the distance that a ball rolls down an incline in each succeeding second. If it rolled 1 foot during the first second, it would roll 3 feet during the next, 5 during the third, 7 during the fourth, and so on. Or, if we prefer to put this in another form, the ball would have rolled 1 foot *at the end* of the first second, 4 feet at the end of the second, 9 feet at the end of the third, 16 feet at the end of the fourth, and so on. The distance that the ball has rolled at the end of any second is equal to the square of the number of seconds. This may be written more briefly

$$\text{Distance} = \text{number of seconds squared}$$

or more compactly

79

$$D = t^2$$

Now it turns out that the distance *in feet* happens to equal the square of the number of *seconds* only because the example was chosen to simplify the discussion. All we need do is tilt the incline a bit more and the ball will obviously fall faster—and the formula will no longer be true. But no matter what angle we choose for the incline, there is an underlying relationship between distance and time that always governs the motion of the ball. If the ball has rolled 1 foot at the end of a *certain length of time,* it will have rolled 4 feet at the end of twice that length of time, 9 feet at the end of 3 times that length of time, 16 feet at the end of 4 times that length of time, and so on.

Time	Distance
1 unit	1 foot
2 units	4 feet
3 units	9 feet
4 units.	16 feet

This rule is expressed in the following mathematical formula:

$$D = \tfrac{1}{2} \times at^2$$

The term *a* in this expression is the quantity known as *acceleration.* It is nothing more than the rate at which the velocity of the ball changes each second. Suppose the incline is removed completely, and the object is allowed to fall freely to earth. The acceleration is then equal to that of gravity, 32.2 feet per second per second. To illustrate,

$$D = \tfrac{1}{2} \times (32.2)t^2 = 16.1 \times t^2$$

At the end of the first second, the object will have fallen 16.1 feet. This means that its *average* speed was 16.1 feet per second. Since it had no speed at all at the instant it was dropped, its terminal speed at the end of the first second must have been 2 × 16.1 = 32.2 feet per second. At the end of the next second the distance fallen was 4 × 16.1 = 64.4 feet. Its average speed for the 2-second time interval was 64.4 ÷ 2 = 32.2 feet per second. Its terminal speed, therefore, must have been twice that amount or 64.4 feet per second. At the end of the third second it had fallen $9 \times 16.1 = 144.9$ feet. Its average speed for the 3 second interval was then $144.9 \div 3 = 48.3$ feet per second, and its terminal speed was 96.6 feet per second.

Time interval	Distance fallen	Terminal speed
1 second	16.1 feet	32.2 feet per second
2 seconds	64.4 feet	64.4 feet per second
3 seconds	144.9 feet	96.6 feet per second
4 seconds	257.6 feet	128.8 feet per second

As you can see, the speed increases by the amount 32.2 during each succeeding second. Or we may say that the speed increases at the rate of 32.2 feet per second each second. This is precisely the meaning of acceleration. It is the rate at which speed changes. It is expressed in feet per second per second and is sometimes written feet/sec/sec.

When scientific principles such as the laws of motion are expressed in mathematical shorthand, we have a formula—a convenient and compact method of conveying a scientific principle.

How many prime numbers are there?

In the *Elements* of Euclid, written some three hundred years before our era, it was shown that the number of primes is infinite. The proof is really quite simple.

Any number is either prime or composite. If it is prime, it has no divisors (if we neglect the uninteresting case of division by 1). If it is composite, it can be expressed as the product of prime numbers. The number 30, for example, is the product of three prime numbers, $2 \times 3 \times 5 = 30$. If we multiply all the known prime numbers we will obtain a very large composite number. Let's call it n for simplicity. Adding 1 to this number gives $n + 1$. But no number, except 1 of course, can divide evenly both n and $n + 1$. So $n + 1$ must be either another prime number or a composite number which has as a factor a prime that was not

81

in the group of primes that we multiplied in the first place. In either event, we have shown another prime to exist. No matter how many primes we may discover, there will be always one more in existence. So, their number must be infinite.

Does a long line contain more points than a short one?

In the study of mathematics, all roads seem, inevitably, to lead to Greece. In evaluating the inability of the ancient Greeks to develop higher forms of mathematics, three major blind spots come to mind: the absence of a notational symbolism, an unspeakable distaste for irrational numbers like the square root of 2, and a horror of the infinite. This *horror infiniti* was a bidirectional myopia, affecting equally the study of numbers that are infinitely large or *infinitely small*. When European thought awakened from the thousand-year stupor of the Dark Ages, the notion of the infinite was one of the first to be attacked. And the person to attack it was one of the greatest thinkers of our millennium, Galileo. In his *Dialogue Concerning the New Sciences*, which appeared in 1632, Galileo presents the first thoughtful study of the infinite. His argument is essentially this: Consider *all* the squares arranged in order, 1, 4, 9, 16, . . . With this in mind, it's possible to count the squares by assigning a number to each; 1 is the first square, 4 is the second, 9 is the third, and so on. No matter how many squares we consider, there will always be a number to go with it. Similarly, no matter how many numbers we may have, we can always find its square. The conclusion, in Galileo's words is this: "So far as I see, we can only infer that the number of squares is infinite and the number of their roots is infinite; neither is the number of squares less than the totality of all numbers, nor the latter greater than the former; and finally the attributes 'equal,' 'greater,' and 'less' are not applicable to infinite, but only to finite quantities. When, therefore, Simplicio introduces several lines of different lengths and asks how is it possible that the longer ones do not contain more points than the shorter, I answer him that the one line does not

82

contain more, or less, or just as many, points as another, but that each line contains an infinite number."

Once Galileo had released the minds of men from the shackles of the finite, mathematics literally went wild. The critical rigor of the Greeks was completely forgotten. The infinitely small quantity, or infinitesimal, was introduced and its use broke all the laws of mathematical decorum. Kepler, Cavalieri, Newton, Leibnitz, Wallis, Euler—all these dealt with infinitesimals in a manner that would have caused the most progressive Greek mathematician to shudder. There was no Euclid to keep this mathematical orgy in check and the storm it raised is far from settled yet. But despite their lack of rigor in the classical sense, Newton and Leibnitz developed independently the analysis of the infinitesimal which we know today as the *calculus*—one of the most important tools of science.

Is the number 13 unlucky?

Three possible inferences can be drawn concerning the number 13; either it's *lucky* (which is highly improbable), *unlucky* (which seems the case), or *passive* (which is too uninteresting to warrant consideration). Having made up our minds we can now proceed to prove the diabolical nature of 13 through an irrefutable process known as *argument by enumerating instances*.

During the course of a day many influences touch each of our lives and quite naturally some of these will be unfortunate. But, happily, there's something at hand (other than ourselves) upon which to throw the guilt. Of the many small numbers that surround us, 13 has a history of unfortunate associations and little effort on our part is needed to further blacken its name. Weren't 13 steps involved in our recent rapid and unconventional descent down a flight of stairs? Or, if not, weren't we carrying 13 cans of beer, or 13 apples, or 13 pennies in our pocket? Failing any of these, weren't there 13 panes of glass in the room, or 13 chairs, or 13 persons waiting for our appearance? We can count or measure anything at all and no room, or streetcar, or theater worthy of the name will fail to produce 13 of *something*. Of course, we will have to ignore any misfortune not connected with 13 and, similarly, any good fortune that *is* connected with 13. In this way, we can prove conclusively that the reputation of 13 as a bad actor is truly based on fact. Thus the enumeration of many instances will prove to the most confirmed skeptic that 13 is indeed an unlucky number.

The technique described above can also be applied to foretelling the future. Notice that fortunetellers and other mystics always present their portents in general terms, as if their supernatural vision were clouded somewhat by the great physical strain involved in such exertions. Typically, a mystic will notice in his crystal ball a stock certificate of a powerful corporation. Unfortunately, the distance is too great to permit reading the name of the company (or even the New York Stock Exchange symbol, worse luck). But the vision is clear enough to show the stock moving up a steep incline (obviously an indication of increasing value) while the certificate divides in half—amoeba fashion—to produce two new certificates (obviously a stock split). When pressed hysterically for an identification, the mystic replies that the best he can do is to note that each certificate seems to be carrying a load of metallic objects, pointed on one end and blunt on the other. And then the image goes dead. Naturally, the clients rush out to buy National Missile Makers, Inc., in the expectation of making a huge profit. It turns out, 13 days later, however, that the American Nail Company announces a 4-for-1 stock split and goes up 38 points! Such are the vicissitudes of the clairvoyance business.

Whenever argument by enumeration is substituted for a decent respect for the long arm of coincidence, there is likely to be a whopper of a wrong conclusion just around the corner. Is 13 an unlucky number? On the basis of the evidence, the reputation of 13 is probably no more significant than that of any other number—with the possible exception of 7, which is a particularly lucky number, as everyone knows.

Can we be sure that the earth rotates?

In 1851, the French physicist Foucault first demonstrated a convincing proof that the earth really does rotate on its axis. Foucault suspended a heavy iron ball from the dome of the Pantheon in Paris by a wire more than 200 feet long, and started his pendulum swinging (Figure 18). Those present saw the plane of the pendulum's movement slowly turn in a clockwise direction. To understand the operation of Foucault's pendulum, imagine such a pendulum swinging directly over the North Pole. The earth rotates 15° in 1 hour, but the pendulum's absolute direction remains unchanged. Since the earth rotates from west to east, the pendulum seems to rotate 15° per hour in a clockwise direction.

The rate at which the pendulum's plane of motion changes depends on the latitude of its location. At the pole it changes 15° in 1 hour; at the equator there is no change at all; at the latitude of Chicago the rate of change is 10° per hour.

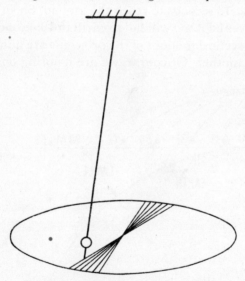

FIGURE 18. The Foucault pendulum, a demonstration of the earth's rotation.

What keeps an artificial satellite up in the sky?

No one knows what gravity is—or, in fact, whether or not there *really* is such a thing as gravity at all. The theory of relativity does quite nicely without it. But of one thing we are certain: everyday objects near the surface of the earth behave *as if there were* a force of gravity. Drop an object and it falls. Throw a ball up into the sky and it returns to earth. Early mathematicians like Galileo and Newton found order in the way in which objects fall. They discovered that objects don't just fall, they fall in a precise and predictable manner. A stone dropped from a cliff will fall vertically in accordance with a simple equation $s = \frac{1}{2}gt^2$—where s is the distance through which the stone falls in a time interval t, and g is the *acceleration due to gravity*, or 32.2 feet per second per second. If our hypothetical cliff were 1 mile high, it would take the stone about 18 seconds to hit the plain below.

Now it doesn't really matter whether you *drop* the stone from the cliff, or *throw* it out horizontally as fast as you can, it will still take just 18 seconds to reach the ground. Shoot a bullet out horizontally, and it too will fall to earth in 18 seconds. The horizontal and vertical motions of the projectile are quite independent of one another. Of course, we are ignoring one important

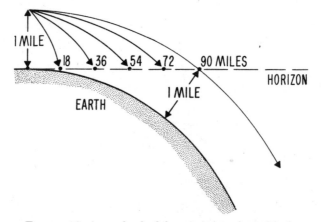

FIGURE 19. A method of determining the orbital velocity of an artificial earth satellite.

86

point in all this. As the speed of the projectile gets greater and greater, we will have to take into account the fact that the earth isn't flat. The earth curves down and away from the horizontal direction as we get farther and farther away from the starting point. A glimpse at Figure 19 will illustrate this point. Suppose we fire a missile horizontally from a point 1 mile above the earth's surface. Let's say that the muzzle velocity is 1 mile per second. The missile will take just 18 seconds to "drop" to the horizon line and in that length of time it will travel horizontally exactly 18 miles from the starting point. But it will take a little longer to reach the earth because of its curvature. If we shoot the missile at 2 miles per second, it will travel 36 miles before reaching the horizon line. At 3 miles per second, 54 miles, and so on. As you can see, the faster the missile travels, the higher it is above the earth's surface at the end of its 18-second "fall." At this point, a question probably suggests itself: At what speed must we fire the missile so that it will remain a constant distance from the earth's surface? A second look at the diagram will help us find the answer. We know that the missile "falls" 1 mile in 18 seconds. We also know that it must end up 1 mile above the surface at the end of this length of time, or it will not continue to circle the earth indefinitely. The question, then, may be turned around a bit: How far must we travel before the earth drops away 1 mile below the horizon? This is the same as inquiring the distance to the horizon at a height of 1 mile. A rule developed in another part of the book tells us that this distance is 90 miles, which, therefore, is the distance that the missile must travel horizontally in 18 seconds. Dividing these figures gives us the *orbital velocity* that our object must attain in order to circle the earth indefinitely—a velocity that turns out to be 5 miles per second or 18,000 miles per hour.

Of course, we have left something else out of our calculations —air resistance—or the resistance of our atmosphere to motion. Even if we did fire our missile at 5 miles per second, it would soon be slowed down by the air and fall to earth. In order to establish a satellite, we must shoot off our missile at heights of

100 miles or more, where only faint traces of the earth's atmosphere can be found. The Sputniks, Explorers, and Vanguards were all placed in orbit at such heights.

Why does the moon have no atmosphere?

Physicists assure us that the molecules of any gas dart about incessantly, bumping into one another much like little boys in a crowded school playground. Their average speed increases as the temperature is raised, and lighter gases have greater velocities than heavier ones. Hydrogen molecules, for example, move faster than a mile per second *on the average* under normal atmospheric conditions. Molecules of the air around us have average speeds somewhat greater than half a mile per second. But these are averages. At any given instant, some molecules are moving one way, some another; some are traveling at great speed, and some slow; indeed, some are even at rest. It isn't too unreasonable to expect, then, that an errant molecule, located high in the atmosphere, might achieve a sufficiently high velocity in the right direction to pop free of the earth's gravity altogether. Whether or not many molecules will leave a planet's atmosphere in this manner depends on the *velocity of escape* of that planet. This velocity is given* as

$$V = R \sqrt{\frac{2g}{R + h}}$$

where R = radius of earth = 3,960 miles
h = height above earth's surface
g = 32.2 feet per second per second, acceleration due to gravity

Such phenomena as the *aurora borealis* are known to exist at altitudes of about 600 miles, so while the air is extremely thin at such altitudes, faint traces of an atmosphere are still to be found. It's a simple matter to calculate the velocity that a molecule must attain in order to escape from such an altitude.

* See, for example, Carl A. Norman, Richard H. Zimmerman, *Introduction to Gas-Turbine and Jet-Propulsion Design*. (New York: Harper & Brothers, 1948)

$$V = 3960 \sqrt{\frac{(2)(32.2)}{(3960 + 600)\, 5280}}$$

$V = 6.5$ miles per second

This velocity is not too different from the escape velocity at the surface of the earth, 7 miles per second. At great heights above the earth's surface, even those molecules traveling considerably faster than average find it almost impossible to pull free of the earth's pull of gravity.

Conditions on the surface of the moon are quite another matter, however. The moon's gravitational pull is about one-seventh that of the earth's and the corresponding escape velocity, for that body is only 1.5 miles per second. This is low enough to permit the faster moving molecules to escape, and once this process begins, a heavenly body soon loses all its atmosphere.

While we have limited our discussion to escaping molecules of a gas, the same principles hold for any object that is to be "shot" free of the earth's pull of gravity. The word "shot" is emphasized in order to make it plain that the object receives all its energy in one installment. It would be theoretically possible, on the other hand, for a spaceship to leave the earth at a lower speed if the lower force of propulsion were maintained for the entire length of the journey. To use an analogy, a person can either jump up an entire flight of stairs or walk up step by step. The latter method is extremely wasteful of energy, however, and for this reason, rockets are usually ballistic devices that receive their push in as short a length of time as possible.

The velocity of escape for each of the major heavenly bodies of the solar system is given in the following table:

Body	Escape velocity (miles per second)
Earth	6.95
Moon	1.49
Sun	384.30
Mercury	1.99
Venus	6.51

Mars	3.22
Jupiter	38.04
Saturn	23.53
Uranus	14.40
Neptune	12.95

What are the characteristics of a right angle?

Euclid defined a *right angle* ($= 90°$) as the gap between a a plumb line and the horizon, which is the same on all sides. Actually, there is really nothing right or wrong about such an angle, but it does have some very special properties. Two right angles, for example, add up to a straight line. So a straight line is nothing more than an angle of 180°. Carrying this use of the word "right" a step further, a *right triangle* is defined as a triangle having one angle of 90°. An ancient method of making a right angle depends on the fact that any triangle with sides of 3, 4, and 5 units of length is a right triangle. Legend has it that the pyramid builders of Egypt laid out a right triangle by tying together three pieces of rope whose lengths were in the ratio 3: 4:5. If this loop is pulled taut and pegged down at the knots, a perfect angle of 90° is formed.

Another unexpected characteristic of right triangles has come down to us as the Pythagorean theorem. In mathematical terms, the square on the longest side (hypotenuse) of a right triangle is equal to the sum of the squares on the other two sides. This bit of knowledge was known to the Babylonians and Egyptians at least five thousand years ago and probably came down to them from the Orient. Using this relationship, we can verify the truth of the 3-4-5 right triangle and also show that any 5-12-13 triangle is right-angled.

$$3^2 + 4^2 = 5^2$$
$$9 + 16 = 25$$

Similarly,

$$5^2 + 12^2 = 13^2$$
$$25 + 144 = 169$$

Another useful characteristic of the right angle is the help it

90

gives us in determining the area of any triangle. We can show that if we are given any right triangle, we can make a rectangle of twice its size. So the area of any right triangle is equal to half the area of such a rectangle. Next, we can show that any triangle at all can be split into two right triangles by drawing a line from the corner of the large angle perpendicular to the opposite side. So we can find the area of any triangle by multiplying the base length by one-half the height. Carrying this a step further, any odd-shaped figure at all can be split into triangles provided it has straight sides. So we can measure the area of any such *polygon* by reducing it to a number of triangles.

The Greeks knew also that the lines joining the ends of a diameter to any spot on a semicircle enclose a right angle. In other words, any triangle inscribed in a semicircle is a right triangle. This gives us a convenient method of making a right angle with only a straight edge and compass.

As you can see, the right angle has many unexpected but significant characteristics. Of these, perhaps the most important of all was discovered in the second century B.C. by Hipparchus, the greatest astronomer of the ancient world. His discovery created trigonometry, the branch of mathematics that was used so ingeniously to chart the heavens and measure the dimensions of the earth. Hipparchus noted that two triangles are *similar* (note the mathematical sense of this term) if the angles of one are equal to the corresponding angles of the other. Although their *size* may be different, they are similar if their angles are the same. Since the sum of the angles of any triangle is equal to 180°, it's really only necessary to show that two angles of one triangle are equal respectively to two angles of the other, for then the third angles must also be equal. Now if we deal with right triangles, we start off with one angle of 90°, so it is only necessary to show that an acute angle of one right triangle equals an acute angle of the other to conclude that the right triangles are similar. The theorem that Hipparchus applied states that if two triangles are similar, the ratio of any two sides of one triangle equals the corresponding ratio of the other. Let's examine the triangles of Figure 20. Triangles ABC and $A'B'C'$ are right triangles. If angle A = angle A', the triangles must be similar. Hipparchus then concludes that the ratio of the side opposite angle A to the hypotenuse (or longest

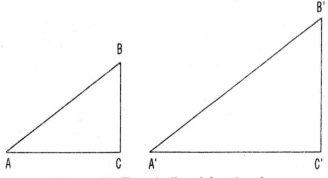

FIGURE 20. Two similar right triangles.

side) must be the same for every right triangle containing angle A. This ratio has become so important that it was given the special name *sine* and the ratio is called by the name sine A.

$$\text{Sine } A = \frac{\text{side opposite angle } A}{\text{hypotenuse}}$$

The same reasoning can be applied to the ratios of the other sides in the right triangle. For example,

$$\text{Cosine } A = \frac{\text{side adjacent to angle } A}{\text{hypotenuse}}$$

$$\text{Tangent } A = \frac{\text{side opposite angle } A}{\text{side adjacent to angle } A}$$

Since these ratios are the same for any right triangles containing angle A, it was only a matter of time before tables were prepared which listed the numerical value of these important ratios for various angles. These tables are used much as we use a dictionary to look up the meaning of a word. When we see sine 34° in an equation, we need only look it up in a table of natural sines to find that it equals 0.5592. These numerical equivalents have been determined once and for all and we need not calculate them each time they are used.

Hipparchus used his marvelous new sines, cosines, and tangents to measure the earth and the heavens without having to lay a

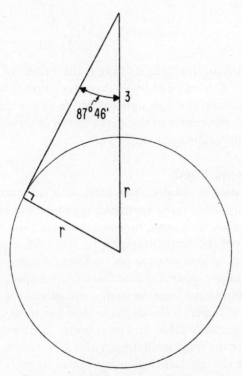

FIGURE 21. Using an ancient idea to
measure the radius of the earth.

yardstick between the two points in question. Here's how we can use his trigonometry to measure the radius of the earth. First, let's measure the height of a mountain. This is relatively easy to do with our new knowledge of trigonometry. Suppose it's 3 miles high. We can then climb the mountain and measure the angle between the mountaintop and the horizon. This angle would be 87° 46′. From Figure 21 we see that

93

$$\sin 87°46' = \frac{r}{r + 3}$$

Hipparchus had already calculated the sine ratios, so he knew that $\sin 87°46'$ is 0.99924 correct to five decimal places. Consequently,

$$0.99924 = \frac{r}{r + 3}$$

Solving this equation tells us that r, the radius of the earth, is equal to 3,944 miles. Although this procedure is somewhat laborious, I'm sure you will agree that it's easier than digging a hole through the center of the earth to the other side and dividing the resulting distance by 2!

What are inequalities?

Mathematics is usually associated with the manipulation of equations in which one expression is placed equal to another. There are many situations, however, in which the nature of the problem forces the mathematician to deal with expressions that are not equal. If you were to pair off your fingers with a dozen eggs, for example, you would be forced to conclude that some of the eggs would be left over. In such a situation we would say that the number of fingers is less than the number of eggs. The symbol $<$ is used to mean *less than*, and its opposite $>$ denotes *greater than*. If F stands for the number of fingers and E for the number of eggs, we can write the inequality in either of two ways

$$F < E$$
or
$$E > F$$

Sometimes one set of objects may be *equal to or less than* another set. We use the symbol $\leq$ to cover that situation and its opposite $\geq$ to denote *equal to or greater than*. In all cases, the larger number is adjacent to the larger portion of the symbol.

Negative numbers are also brought into this scheme. Since -4 is less than -3, we write $-4 < -3$, or $-3 > -4$.

94

If a number n is known to be greater than 4 but less than 7, we write $7 > n > 4$ which is read, "n is greater than 4 but less than 7."

To illustrate the use of inequalities we will prove the theorem which states that *the square of the sum of any two numbers cannot be less than four times their product.* Let's denote any two numbers by a and b. According to the theorem,

$$(a + b)^2 \geq 4ab$$

The inequality is an accurate statement of the theorem since *greater than or equal to* means the same as *cannot be less than*.

To begin the proof, let's assume that a and b are both positive numbers, and then proceed to square the term on the left-hand side, $a + b$.

$$
\begin{array}{r}
a + b \\
\times\ a + b \\
\hline
ab\ + b^2 \\
a^2 + ab \\
\hline
a^2 + 2ab + b^2
\end{array}
$$

Since $(a + b)^2 = a^2 + 2ab + b^2$, the inequality can be written as follows:

$$a^2 + 2ab + b^2 \geq 4ab$$

Subtracting $4ab$ from both sides,

$$a^2 - 2ab + b^2 \geq 0$$

so we must prove that $a^2 - 2ab + b^2$ is positive or at least zero. But it's a fact that that expression is equal to $(a - b)^2$, which we can easily prove.

$$
\begin{array}{r}
a - b \\
\times\ a - b \\
\hline
-\ ab\ + b^2 \\
a^2 - ab \\
\hline
a^2 - 2ab + b^2
\end{array}
$$

The inequality can then be reduced to

$$(a - b)^2 \geq 0$$

But the square of any number, whether it be positive or negative, must be a positive number, so the inequality is true and the theorem is proved if a and b are positive numbers.

If both a and b are negative numbers, the same inequality results since $(-a - b)^2 = (a + b)^2$ and $4(-a)(-b) = 4ab$. So the theorem is also true for that case. If either a or b is negative, the left-hand side is either positive or zero, since the square of any number is positive. But the right-hand side is always negative, so the theorem is proved for this case also.

It has been necessary to write the inequality with $\geq$ instead of $>$ because it's possible that $(a + b)^2$ and $4ab$ may be equal; for example, $(5 \times 5)^2 = 100$ and $4(5 \times 5) = 100$. This will occur whenever the numbers a and b are equal since each side of the inequality then reduces to $4a^2$.

What are continued fractions?

Continued or chain fractions probably aren't too important, but they are different—and surprising. As with many arithmetical principles they seem complicated on the surface, but are reasonable and simple once we understand how they work. In general, an irrational number, like π or $\sqrt{2}$, can be represented by a continued fraction which, in turn, can be approximated by a common fraction. Here's the way it's done.

$$\sqrt{2} = 1 + \sqrt{2} - 1 = 1 + (\sqrt{2} - 1)$$

Multiply $(\sqrt{2} - 1)$ by $\dfrac{\sqrt{2} + 1}{\sqrt{2} + 1} \; (= 1)$:

$$\sqrt{2} = 1 + \frac{(\sqrt{2} - 1)(\sqrt{2} + 1)}{(\sqrt{2} + 1)} = 1 + \frac{2 - \sqrt{2} + \sqrt{2} - 1}{\sqrt{2} + 1}$$

$$\sqrt{2} = 1 + \frac{1}{\sqrt{2} + 1} = 1 + \frac{1}{2 + \sqrt{2} - 1} \tag{1}$$

Multiply $(\sqrt{2} - 1)$ by $\dfrac{\sqrt{2} + 1}{\sqrt{2} + 1}$ once more:

$$\sqrt{2} = 1 + \cfrac{1}{2 + \cfrac{(\sqrt{2} - 1)(\sqrt{2} + 1)}{\sqrt{2} + 1}} = 1 + \cfrac{1}{2 + \cfrac{2 - \sqrt{2} + \sqrt{2} - 1}{\sqrt{2} + 1}}$$

$$\sqrt{2} = 1 + \cfrac{1}{2 + \cfrac{1}{\sqrt{2} + 1}}$$

By continually setting $\sqrt{2} + 1 = 2 + \sqrt{2} - 1$, and multiplying by $\dfrac{\sqrt{2} + 1}{\sqrt{2} + 1}$, we get

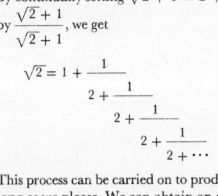

$$\sqrt{2} = 1 + \cfrac{1}{2 + \cfrac{1}{2 + \cfrac{1}{2 + \cfrac{1}{2 + \cdots}}}}$$

This process can be carried on to produce a continuing fraction as long as we please. We can obtain an approximate common fraction to any desired accuracy merely by considering a sufficient number of 2's, discarding the final $(\sqrt{2} - 1)$ and converting the resulting fraction to simplest terms. If, for example, we wanted to consider four 2's in the expression given above, we would get successively

$$\sqrt{2} = 1 + \cfrac{1}{2 + \cfrac{1}{2 + \cfrac{1}{\cfrac{5}{2}}}}$$

97

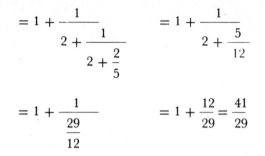

$$= 1 + \cfrac{1}{2 + \cfrac{1}{2 + \cfrac{2}{5}}} \qquad = 1 + \cfrac{1}{2 + \cfrac{5}{12}}$$

$$= 1 + \cfrac{1}{\cfrac{29}{12}} \qquad = 1 + \cfrac{12}{29} = \cfrac{41}{29}$$

The fraction $\frac{41}{29}$ is equal to $1.413793\cdots$ and is an approximation of $\sqrt{2}$ $(1.4142134\cdots)$, correct to three decimal places.

If, like the author, you enjoy doing this sort of thing, perhaps you will want to prove the following identities the next time you're waiting for the menu, or sitting in the dentist's office.

$$\sqrt{5} = 2 + \cfrac{1}{4 + \cfrac{1}{4 + \cfrac{1}{4 + \cdots}}}$$

$$\sqrt{7} = 2 + \cfrac{3}{4 + \cfrac{3}{4 + \cfrac{3}{4 + \cdots}}}$$

$$\sqrt{11} = 3 + \cfrac{2}{6 + \cfrac{2}{6 + \cfrac{2}{6 + \cdots}}}$$

Is there an easy way to add large columns of numbers?

One of the casualties of our business-machine age is the book-keeper of a few decades past who sat on a high stool, wore a green eyeshade, and could do lightning-fast arithmetic in his head. A favorite pastime of these men was simultaneous three-column addition.

$$
\begin{array}{r}
123 \\
654 \\
\underline{987} \\
1{,}764
\end{array}
$$

This is an easy and rapid way to add three columns of figures at the same time. Merely take the top number (123) and add the units figure of the next lower number (4) to get 127. Then add the tens figure (5, which is really 50) to 127, giving 177. Next add the hundreds figure (6, or really 600) to 177, which yields 777. Then add the figure in the units column of the bottom row (7) to get 784; the tens figure (8) to get 864; and the hundreds figure (9) to get 1,764. Once you have learned this trick, you can run down such a column of figures, noting the successive sums as you go: 123, 127, 177, 777, 784, 864, and 1,764. Of course, you can run up as well as down just as readily and the sum will be just as correct.

Writing down such a sum with little apparent effort never ceases to amaze the uninitiated. You can use a bit of trickery, however, to produce an even greater effect. Ask a friend to write a number having about five or six digits and then you write a similar number directly underneath it. Be sure, however, that your number and his are related as follows:

369,247   (friend's number)
630,753   (your number)

Starting with the left-hand side (to look casual) select each digit so that when added to the one above, the sum is 9. Do this until you reach the last digit and select this one to yield 10. (As you can see, this process gives 1,000,000). Then have your friend write another number and follow the same technique with your number. Repeat this process several times and when your last turn comes up, write down *any figures* at all so long as the sums are less than 9.

99

$$
\begin{array}{r}
369,247 \\
630,753 \\
473,291 \\
526,709 \\
974,531 \\
25,469 \\
425,823 \\
\underline{123,022} \\
3,548,845
\end{array}
$$

To add the column start at the left and put down a 3, since there are 3 pairs of 1,000,000's above. Then simply add the last two rows, setting down the figures in their proper order. This trick will go over very well if you don't repeat it too often.

What is the difference between the focal length and the f-number of a lens?

The focal length of a lens determines the size of the image it can produce: the longer the focal length, the larger the image. Focal length and image size are directly proportional—doubling the focal length produces an image that is twice as large. The focal length of a lens is the distance measured from the center of the lens to the image of an object that is a great distance away—the sun, for example (Figure 22). To estimate the focal length of a lens, merely hold it toward the sun and measure the distance at which the sharpest, hottest image is produced. (Never

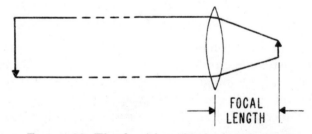

FOCAL
LENGTH

FIGURE 22. The focal length of a lens is the distance measured from the center of the lens to the image of an object that is a great distance away.

100

look at the sun through a lens, however, as severe eye damage may result!) The lens center is more accurately known as the *node of emission* since in certain telephoto and wide-angle lenses this point is located outside the lens proper. The focal length of a lens is usually expressed in inches or centimeters, and it is normally engraved on the lens mount.

For camera lenses, the focal length is the shortest distance between lens and negative that will produce a sharp image. When a camera is adjusted in this way, it is focused on "infinity." To produce a sharp image on closer objects, the lens-to-negative distance must be increased. To produce an image that is the same size as the object, this distance must be approximately twice the focal length of the lens. To produce an image that is twice as large as the object, the lens-to-negative distance must be three times the focal length of the lens, and so on.

The *f*-number of a lens is known technically as its *relative aperture*. It's a measure of the light transmission of a lens. The larger the diameter of a lens in proportion to its focal length, the larger its relative aperture. Since the relative aperture depends upon two factors, focal length and diameter, it cannot be expressed directly as can focal length. Relative aperture equals the focal length divided by the diameter of the front surface of the lens. This ratio is expressed in *f*-numbers.

$$\text{Relative aperture} = \frac{\text{focal length}}{\text{diameter}}$$

If the focal length of a lens is 150 millimeters, and its diameter is 23.5 millimeters, its relative aperture is $150 \div 23.5 = 6.3$. This number is usually written as either *f*.6.3 or *f*/6.3.

The relative aperture of a lens can be adjusted within reasonable limits in practical cameras by means of a movable iris or diaphragm. This element of the lens is comparable to the iris in the human eye in that the useful portion of the lens can be changed from a pinhole at the center, to the maximum diameter of the lens. If a relatively great amount of light is desired to strike the negative, the iris is opened wide. Conversely, when only a small amount of light is needed, the iris is partially closed.

Many beginners in photography find it hard to understand why the *fastest* or most sensitive lenses have the *smallest f*-numbers. Perhaps an example will help to make this point clear. The lens of a camera can be thought of as a window through which light enters the camera. This window has an area determined by the setting of the iris. The larger the window, the greater the amount of light that enters the camera. But this amount of light, whatever it may be, must cover the entire area of the negative. If the focal length is great, the image size is large, and the available light must cover a relatively large area. The light per unit area, or the light intensity on the negative material, will then be relatively low. If, on the other hand, the focal length is short, the image will be smaller, and the light intensity on the negative will be great. From this we can see that the "speed" of a lens must include two factors: focal length and lens diameter. The relative aperture, or *f*-number, is the ratio of these quantities. A small *f*-number, generally speaking, implies that a lens has a relatively short focal length and a relatively large diameter. Both these qualities result in relatively high intensity of light on the negative. Such a lens, therefore, is spoken of as being "fast," since its great sensitivity to light will permit the proper exposure of negatives in a short length of time. The lens can be made slower by partially closing its iris opening. This reduces the effective diameter of the lens without changing the focal length, so the *f*-number is increased. The lever that adjusts the effective opening of a camera lens is calibrated in such a way that each consecutive *f*-number requires twice the exposure of the preceding larger diaphragm opening (i.e., smaller *f*-number). In other words, as the adjustment is moved from one *f*-number to the next lower, the amount of light striking the film is doubled. All things being equal, the length of time that the shutter is open would then have to be halved in order to maintain the same exposure. A lower *f*-number requires a faster shutter speed, so a lens having a relatively low *f*-number is said to be fast.

How was the composition of the sun determined?

Although most of us think of *white* and *colorless* as synonymous terms, it's a fact that white light—light from the sun, for example—is really a mixture of light of all colors. This unexpected property of light was first demonstrated in 1666 by the great English mathematician and scientist, Sir Isaac Newton. Newton allowed a narrow beam of light to enter a darkened room and placed a triangular glass prism in its path. He found that the beam was not only *refracted* (bent) from its original path by the prism, but that it was also spread out into a band of colors. The band exhibited a continuous range of colors from red at the least refracted end of the spectrum to violet at the other. Newton's band of colored light is called a *visible spectrum* and the separation of the colors is called *dispersion*. With his simple experiment, Newton had established an area of scientific investigation that was to become a new and important branch of physics—a discipline known as spectroscopy, the analysis of light.

Spectroscopy has been developed to such an extent during the past century that, through it, scientists can study the temperature, chemical make-up, and motion of any light source, whether it be a minute flame on earth or the radiation of a distant galaxy. As we shall see, the principles of spectroscopy have been applied even to the signals radiated by unseen "radio stars," making possible the study of heavenly bodies that are too faint to be seen or are hidden forever from visual observation by obscuring dust clouds.

In 1802, W. H. Wollaston, in England, improved upon Newton's equipment by causing the beam of light to go through *two* slits instead of one. He reasoned that a source of light having a large angular size like the sun would allow the rays from different parts of the source to strike the prism at different angles. This would mix the colors of the resulting spectrum somewhat; the green light from one point on the sun, for example, falling on the red light from another. Having improved the parallelism of the rays, Wollaston noticed that the spectrum of sunlight was crossed in a number of places by dark lines which did not occur in the light of a candle flame. A few years later, the great German telescope maker, Fraunhofer, improved the spectroscope further by using a single narrow slit, a lens to make the rays more parallel, and a telescopic eyepiece. With his magnificent instruments, Fraunhofer found about seven hundred dark lines and although he could not explain their true meaning, he realized their importance as points of reference. He mapped their positions carefully with letters of the alphabet, beginning in the red with *A*. Such lines are now known as Fraunhofer lines. A spectrum containing them is a *dark-line spectrum.*

The next important step was taken in England in 1822, when Sir John Herschel noticed that a flame impregnated with a certain substance would produce a spectrum having isolated *bright* lines. And significantly, the position of these bright lines was *characteristic of the substance with which the flame was fed!* The chemical composition of an unknown substance could be analyzed spectroscopically to determine its chemical composition! This kind of spectrum is known as a *bright-line spectrum.*

Back in Germany, Fraunhofer found the clue that united the bright- and dark-line spectra. He and others noticed that the bright yellow line of the sodium bright-line spectrum was *identical* in position to the dark line in the spectrum of the sun, which he had lettered *D*. Later, Kirchhoff and Bunsen performed additional experiments which formed the presently accepted principles of spectroscopy.

1. An incandescent solid or liquid gives a *continuous* spectrum consisting of all colors unbroken by dark or bright lines. (Gases under great pressure, such as at the center of the sun, also produce continuous spectra.)

2. An incandescent gas under relatively low pressure gives a bright-line spectrum, the positions of the lines being characteristic of the chemical composition of the gas. Bright-line spectra are also known as *emission spectra,* since the lines are *emitted* by the gas. All elements produce at least two bright lines and some of the metals (in gaseous form) produce hundreds or thousands of bright lines. No two elements are known to have a single line in common.

3. If light that would produce a continuous spectrum passes

through a gas that is *cooler* than the light source, dark lines appear in the spectrum in the positions characteristic of the bright lines belonging to the gas. In other words, a gas *absorbs* just those rays that it is capable of emitting—no others. For this reason, dark-line spectra are often referred to as *absorption spectra.*

By comparing the dark-line spectrum of sunlight with the spectra of terrestrial elements, it has been possible to determine that some 65 chemical elements exist in the atmosphere of the sun. Of these, hydrogen and helium are most abundant.

Thus far, the lines of a spectrum have been labeled by their color or by an arbitrary letter, such as was assigned by Fraunhofer. In studying spectral lines quantitatively, it's necessary to become more specific.

It's well known that light and radio radiations are merely different portions of a continuous *electromagnetic spectrum.* Although no one really knows what they *are,* scientists do know that they behave much like water waves; by this is meant they exhibit a wavelike *behavior.* The length of a water wave is the distance between successive peaks, or troughs. It can be measured with a yardstick. In much the same way, the length of radio and light waves can be expressed in meters, or centimeters ($\frac{1}{100}$ meter), or millimeters ($\frac{1}{1000}$ meter). The wavelength of red light, for example, is about 0.000078 centimeters. Since the velocity of light is approximately 3×10^{10} centimeters per second, the number of red waves passing a given point in 1 second is $3 \times 10^{10} \div 0.000078 = 385 \times 10^{12}$ per second, which is the *frequency* of red light. The usual symbol for wavelength is λ (lambda), so the frequency f can be expressed

$$f = \frac{c}{\lambda}$$

where c is the velocity of light.

When astronomers photograph the spectra of nebulae far beyond our own galaxy, they find all the familiar spectral lines to be displaced toward the red end of the spectrum. In other words, the wavelengths of these lines are *longer* than they "ought" to be. The magnitude of this *red shift* depends upon the distance of the nebulae. The greater the distance, the greater the shift in wavelength of the lines. The Doppler-Fizeau effect, as it is known, is analogous to the reduction in pitch of a receding sound source. In 1848, Fizeau formulated the principle which is of incalculable importance in astronomy: When a light source recedes from an observer, the lines of its spectrum lie farther to the red by an amount proportional to its velocity. In algebraic form

$$\text{Change in wavelength} = \lambda \frac{v}{c}$$

where v = velocity of recession
       c = velocity of light
       λ = wavelength of spectral line under study

By this means, it has been determined that some distant nebulae are receding from our galaxy at speeds equal to a large fraction of the speed of light!

Perhaps even more amazing is the recent discovery that spectral lines can extend into the relatively long wavelength region occupied by radio waves! Hydrogen, the building block of stars and galaxies, has associated with it a spectral line whose wavelength is 21 centimeters long—just about one-half million times longer than those of visible light. In April of 1944, H. C. van de Hulst, the Dutch astronomer, predicted the existence of the hydrogen radio line as emission from the hydrogen atom. In 1951, Ewen and Purcell of Harvard were able to detect the 21-centimeter radiation. So important was this discovery that it was immediately verified by Muller and Ort in the Netherlands and Pawsey in Australia. Since that time, the hydrogen radio line has become the most studied portion of the spectrum. Powerful radio receivers with large antennas are used to detect this wavelength from various points in the sky. In 1955, Lilley and McClain at the United States Naval Research Laboratory reported the red shift of the 21-centimeter radio line in connection with the second most intense radio star. This was the Cygnus A "radio star" at a distance of 10^8 light-years. Using the 200-inch

106

Hale telescope, Baade and Minkowski had previously discovered that this "star" was really the collision of two distant galaxies. It was subsequently determined that the optical red shift corresponds closely to the shift observed in the wavelength of the 21-centimeter hydrogen line. For the first time, the optical and radio "red shifts" had been measured on the same object and found to be in substantial agreement.

The importance of this experiment can hardly be overestimated. It's well known that the magnitude of the observed red shift is proportional to the distance of the emitting nebulae. The greater the distance, the greater the change in wavelength. If the red shift is actually a result of the Doppler-Fizeau effect, it would seem that all distant galaxies and nebulae are receding from the earth and from each other—in other words, the universe is expanding! The recent 21-centimeter experiments seem to bear out these conclusions. Radio astronomers expect that the hydrogen line and others still to be found will provide tools as powerful as have been the visible spectra in optical astronomy.

How is the height of a mountain measured?

If you have a method of measuring angles, and a table of natural tangents, it's a simple matter to measure the height of any object even though you may not be able to get close to the base. All you have to do is sight the angle of the top of the object, then walk *any* measured distance directly toward or away from

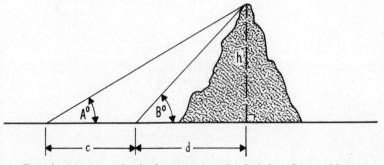

FIGURE 23. A method of measuring the height of any object.

107

the object, and measure a second angle. Figure 23 shows the important angles and distances of the measurement. Angles A and B and distance c are the measured quantities. Distance h is the height that we are looking for. You will recall that the tangent of any acute angle of a right triangle is equal to the length of the side opposite to the side adjacent to the angle. (The expression *tan-o-pad* somehow helped us to remember this relationship in high school.) We can then write expressions for the tangents of angles A and B.

$$\tan A = \frac{h}{c+d} \qquad \text{or } h = (c + d) \tan A \qquad (1)$$

$$\tan B = \frac{h}{d} \qquad \text{or } d = \frac{h}{\tan B} \qquad (2)$$

Combining (1) and (2), we get

$$h = \left(c + \frac{h}{\tan B} \right) \tan A$$

$$h = c \tan A + h \frac{\tan A}{\tan B}$$

$$h - h \frac{\tan A}{\tan B} = c \tan A$$

$$h \left(1 - \frac{\tan A}{\tan B} \right) = c \tan A$$

$$h = \frac{c \tan A}{1 - \frac{\tan A}{\tan B}}$$

$$h = \frac{c \tan A}{\frac{\tan B - \tan A}{\tan B}}$$

$$h = \frac{c \tan A \tan B}{\tan B - \tan A} \qquad (3)$$

Equation (3) gives the height of the object, h, in terms of quantities that are known—the two angles and the distance between the points of observation. Now we can assume numbers for A, B, and c and use equation (3) to solve any such problem. Suppose that you measure angles to the top of a mountain of 35° and 50° from two points on the ground separated by 2,000 feet. Then, from the table of tangents

$$\tan A = \tan 35° = 0.7002$$
$$\tan B = \tan 50° = 1.1918$$

In addition, the problem states that distance c is 2,000 feet. Substituting these numbers in equation (3) gives

$$h = \frac{(2,000)(0.7002)(1.1918)}{1.1918 - 0.7002} = 3,395 \text{ feet}$$

As you have seen, the steps leading to equation (3) were completely general in that letters were used instead of numbers. It may be argued that the solution would have been clearer if we had started with numbers for the known quantities. This would have given the answer directly, since we would have performed the various arithmetical operations as we went along. The reason we didn't follow such a procedure is that equation (3) has solved *all problems* of this kind, once and for all. Equation (3) is a *formula* that can be applied to the measurement of any mountain, building, telephone pole, or obelisk in existence, anywhere that you may care to go. With the aid of this formula, determining the height of any object has been reduced to a problem in arithmetic.

What is the difference between markup and discount?

The markup M on a product is the amount by which the advertised or list price L exceeds the cost C, and the discount D is the amount by which the list price exceeds the actual selling price S. Markups are always calculated on cost, and discounts are always calculated on list price. If a certain item costs $1 and is list priced at $1.40, the markup is $1.40 − $1 = $0.40. If the product is then sold for $1.30, the discount on the sale amounts to $1.40 − $1.30 = $0.10. In general,

109

$$M = L - C$$
and
$$D = L - S$$

It's the usual practice for the seller to compute his gross profit P as a percentage of total sales. In the example given above, the profit was the difference between the selling price and the cost, or $1.30 − $1 = $0.30. Expressed as a decimal, the rate of profit p was equal to

$$p = \frac{P}{S} = \frac{0.30}{1.30} = 0.23$$

or 23 per cent of the selling price. Similarly, the markup rate m is equal to

$$m = \frac{P}{C} = \frac{0.30}{1.00} = 0.30$$

or 30 per cent of the cost.

A problem facing businessmen is to determine the markup rate that must be used in order to provide a certain rate of gross profit on sales. If m and p are defined as above,

$$m = \frac{p}{1 - p}$$

If a retailer desires to realize a gross profit of 40 per cent on sales,

$$m = \frac{0.4}{1 - 0.4} = \frac{0.4}{0.6} = 0.67$$

so he must apply a markup of 67 per cent. Notice that he can't afford to give a discount of 67 per cent on the sale of merchandise. If an item costs $1 and is list-priced at $1.67, a discount of 67 per cent would amount to $1.11, and the selling price would be $1.67 − $1.11 = $0.56, which is considerably less than cost.

It has become popular lately, for psychological reasons, to inflate one's list price so that large discounts may be offered to the buyer. This is not at all difficult to do. Suppose a seller wants to realize a rate of profit on sales equal to p while offering a discount rate d. This can be done quite easily by using *two* markups. The first markup m_1 is calculated as above and the selling price established as follows:

$$\text{Selling price} = \text{profit} \quad + \text{cost}$$
$$S = P \qquad\quad + \ C$$
$$S = \frac{P}{C} \times C + \ C$$
$$S = m_1 \times C + \ C$$
$$S = C(m_1 + 1)$$

A second markup m_2 is then calculated such that

$$m_2 = \frac{d}{1 - d}$$

The list price is then equal to

$$L = C(m_1 + 1)(m_2 + 1)$$

by reasoning analogous to that given above. The formula may look forbidding, but it's not too difficult to use—and it really does wonders for sales. According to newspaper accounts, some manufacturers have employed this technique to the point of printing many different price tags for the same article, thus enabling the retailer to select the one most nearly in agreement with his own particular discount policy. It's beyond our scope to comment on the ethics of such practices except to note: *caveat emptor,* let the buyer beware!

What are repeating decimals?

The decimal $1.4142135\cdots$ is the well-known decimal numeral for $\sqrt{2}$. No matter how far we carry on the calculation of this obstinate number, we never come to an end. Irrational numbers like $\sqrt{2}$ can't be expressed by a digit or by a ratio of digits. The best we can do with whole numbers is to approximate their numerical value.

Decimals that repeat themselves, however, are quite another thing. Let's take the decimal $1.3658536585\cdots$ for example. This *periodic* decimal contains a group of five digits that repeat—36585.

111

Because of their periodic nature, decimals of this kind can be converted into common fractions.*

$$\mathcal{N} = 1.3658536585\cdots$$
$$100,000\mathcal{N} = 136,585.36585\cdots$$

Subtracting $\mathcal{N}$, $\quad\quad \mathcal{N} = \quad\quad 1.36585\cdots$

$$\overline{99,999\mathcal{N} = 136,584.00000000000}$$

$$\mathcal{N} = \frac{136,584}{99,999} = 1\frac{36,585}{99,999}$$

$$\mathcal{N} = 1\frac{45}{123}$$

The converse of this procedure is also true; that is, any common fraction can be changed into a repeating decimal by division.

The number 142,857 seems innocent enough, but it will put on a rather spectacular performance with a little help. Multiply it by any digit less than 7 and the result is a *cyclic permutation* of the original number. The order of the digits remains the same, but the sequence begins with a different digit.

$$1 \times 142,857 = 142,857$$
$$2 \times 142,857 = 285,714$$
$$3 \times 142,857 = 428,571$$
$$4 \times 142,857 = 571,428$$
$$5 \times 142,857 = 714,285$$
$$6 \times 142,857 = 857,142$$

Of course, no one will be satisfied at this point without finding out what happens when 142,857 is multiplied by 7.

$$7 \times 142,857 = 999,999$$

* Multiply the number by 10^a where a is the number of digits in the repeating portion of the number; subtract the number; then simplify.

112

Carrying all this one step further: $\frac{1}{7} = ?$

```
        0.1428571
      _____
    7)1.0000000 ···
      7
      __
      30
      28
      __
       20
       14
       __
        60
        56
        __
         40
         35
         __
          50
          49
          __
           10
            7
           __
            3    (the first remainder above)
```

The first six steps yield different remainders, 3–2–6–4–5–1, but the seventh remainder is 3—the same as the first. The seventh division is 7 into 10, which is exactly what we started with. The entire process then begins to repeat over and over again and *has* to become periodic.

The fraction $\frac{1}{7}$, as we have seen, decimalizes into 0.142857 ···, a repeating decimal. How about $\frac{2}{7}$, which is $2 \times \frac{1}{7}$? As you have probably guessed, it starts out with a different digit, but soon looks just like $\frac{1}{7}$. The same holds true for $\frac{3}{7}$, $\frac{4}{7}$, $\frac{5}{7}$, and $\frac{6}{7}$. Even $\frac{7}{7}$ works out well enough, since

$$7 \times 0.142857 \cdots = 0.999999 \cdots = 1.$$

If you're interested in finding other fractions that behave in this way, look for fractions $\frac{1}{a}$, whose decimal equivalents have a repeating decimal consisting of $a - 1$ digits, the maximum possible period. The only other numbers under 100 demonstrating this quality are 17, 19, 23, 29, 47, 59, 61, and 97. There is no known easy way to find numbers of this kind.

113

What is a googol?

According to Edward Kasner, co-author with James Newman of *Mathematics and the Imagination*, the word "googol" was invented by his nine-year-old nephew to designate the large number 1×10^{100}. Strangely, the number seems to have caught on and is being used more and more frequently as the need for great numbers increases.

At first glance the number 1 followed by 100 zeros would seem ordinary enough, and while it may be quite large, we would expect it to pop up quite often in our modern world of superlatives. But a little investigation in this area yields surprising results. Suppose the earth were made entirely of sand. How many googols do you suppose the total number of grains would make? Ten, a hundred, a million? Actually, all the grains of sand needed to fill a sphere as large as the earth would amount only to about 1×10^{32}—a large number, but submicroscopic when compared to a googol! But grains of sand are relatively large compared to molecules; suppose we were to count all the molecules in the earth. That total would be somewhat larger, having perhaps 50 digits. But even that number fails to approach 10 to the power of 100.

To get into the same league with googols, we must leave the world of physical things (because there just aren't enough of them) and enter the world of electronics. Take the transatlantic telephone cable for example. Entering the ocean at Clarenville, Newfoundland, it stretches 2,250 miles under the North Atlantic until it emerges again at Oban, Scotland. For the greater part of its length it lies in water only a few degrees above freezing on the primeval ooze a mile or two beneath the surface. And distributed somewhat evenly along its length are 51 electronic amplifiers (often called repeaters) which keep the voice impulses from fading away to an undetectably low level. Since there are two such cables, one carrying westbound conversations and the other eastbound, there are really 102 repeaters in the complete transatlantic link. Those pieces of electronic equipment form the heart of the system, for without them the cable could not possibly carry voices across the ocean.

The repeaters are spaced some 40 miles apart. In order to make up for the losses in the 40 miles leading up to it, each repeater must amplify, or increase, the strength of the incoming signals approximately a million times. Since there are 51 repeaters in a one-way cable, the total amplification in each circuit reaches the staggering figure of *a million multiplied by itself fifty-one times!* Just contemplate 10^{306}! That number is so large that even the mighty googol shrinks from its presence in awe! The number 10^{306} is equal to a million googols cubed, or $10^6 \times (10^{100})^3$. That, you will agree, is quite a lot of amplification.

How can shadows be used to measure an object's height?

By the year 300 B.C., Euclid had written 12 books on mathematics. Some 1,700 years later, however, the syllabus of the University of Oxford took the student only to the fifth proposition of

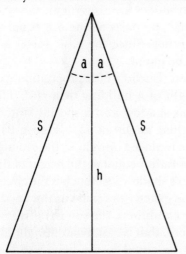

FIGURE 24. The *pons asinorum.*

Euclid's first book. This proposition tells us that *the base angles of an isosceles triangle are equal.* Put in another form, it states that if two sides of a triangle are equal (an isosceles triangle), then the angles

115

opposite these sides must be equal. Since this proposition marked the limit of mathematical achievement at medieval universities, it was referred to as the *pons asinorum,* or *bridge of asses,* because fools couldn't understand the proof of this theorem and, like asses at a bridge, could proceed no further. The proof ran something like this: pick out the angle that is formed by the two sides that are known to be equal (Figure 24). Bisect this angle. We now have two triangles, and the line just drawn is common to both. In addition, the bisected angle has formed two identical angles (by definition), one in each triangle. Finally, the two equal legs of the original isosceles triangle are known to be equal (again, by definition). So we have two triangles—two sides and the included angle of one are equal to two sides and the included angle of the other. From a previous theorem of Euclid's first book, such triangles are known to be *congruent,* or identical. (Euclid dismissed the fact that one triangle is just the reverse of the other as an incidental nuisance, since one could be turned over to fit exactly on top of the other.) Since the triangles are congruent, the very definition of the term tells us that the parts of one are equal to the corresponding parts of the other. Since the base angles are corresponding parts, they must be equal.

Aside from its historical interest, Euclid's fifth proposition can be useful in measuring the height of a building or a cliff. The technique makes use of the sun's shadow and a special kind of isosceles triangle having two equal angles of 45° (Figure 25). Here's how it's done. A short pole is placed upright in the ground. A circle is drawn around it with a radius equal to the height of the pole above ground. When the sun's shadow just touches the circle, the length of the shadow must be exactly the same as the height of the shadow pole. Under these conditions, the sun is 45° above the horizon. Any shadow, therefore, that is cast by an upright object like a cliff or building will be equal to the height of that object. This height can be determined by pacing off the length of the shadow.

This method will not work if the point directly under the peak of the building is inaccessible. We have to measure the total length

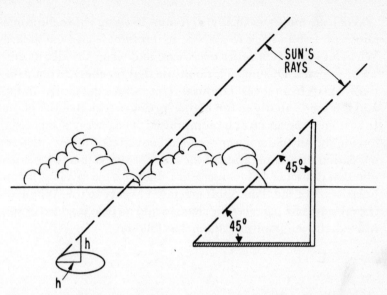

FIGURE 25. Using shadows to measure the height of tall objects.

of the shadow from its tip to the point directly under the spot whose height is to be measured. If the physical limitations of the situation prevent making this measurement, we must resort to the more complicated methods of trigonometry. But having crossed the *pons asinorum,* we will solve that problem in another question.

What is the cause of a chinook?

The *chinook* of the Western plains of North America is a freak wind of warm air that arrives suddenly in the dead of winter, raises the temperature 30° or 40°F in the space of a few hours, and just as suddenly disappears. In the northern Alps, these awesome winds are called the *foehn,* and in Southern California, the *Santa Ana.* Meteorologists believe that chinooks result when air from a plateau or mountain area is brought down vigorously into an adjoining plains region. During the descent, the relative humidity of the air falls rapidly so that the air becomes extremely dry.

117

When air moves vertically, as it must in rising above the mountains and falling back to the sea, its pressure is affected greatly. When air rises, its pressure decreases; and when air falls, its pressure increases. During a chinook, air that reaches the plains actually starts from several thousand feet above the mountaintops and is forced rapidly to the higher-pressure region of the plains. In so doing, the air must be compressed. The energy dissipated in forcing the air to the lower region is converted into heat, thereby increasing the temperature of the air. The derivation of the equation that describes this temperature change is too involved for our purposes, but the final result is relatively easy to use. Whenever the pressure of a parcel of air changes and no heat is added or subtracted, the temperature changes as follows:

$$\frac{T}{t} = \frac{P^{0,288}}{p}$$

Where T is the absolute temperature at the initial pressure P, and t is the absolute temperature at the final pressure p. This equation is often called *Poisson's equation*.

One example of a chinook at Havre, Montana, has been recorded in which the temperature rose 33°F in about an hour when the chinook arrived on February 15, 1948. During another, the temperature at Denver, Colorado, rose 25°F in about two hours shortly after midnight on January 27, 1940. Other local names for chinook-type winds are the *zonda* of Argentina and the *Koembang* of Java.

Who first measured the speed of light?

The speed of light is, perhaps, one of the most important constants found in nature. It is one of the very few things, for example, that we know about empty space. It represents an upper limit, according to the theory of relativity, for the velocity of any material object. No radio energy may exceed it. And in addition to all this, it enters into the equation concerning the equivalence of mass and energy,

118

$$\text{Energy} = \text{mass} \times c^2$$

No wonder men have attempted through hundreds of years to measure it to the greatest possible accuracy. The first such attempt was made by the astronomer Roemer in 1676.

When we consider the fantastic speed at which light travels, it's a wonder that anyone thought to measure it at all. Perhaps they wouldn't have for generations if Roemer hadn't noticed something odd about the eclipses of a moon of Jupiter. Jupiter has several satellites, and Roemer was observing the one nearest the planet. This moon happens to have an orbit that lies approximately in the plane of Jupiter's orbit around the sun. This means that the moon enters the shadow of the planet once each revolution, as shown in the diagram (Figure 26). On the average, this happens once every 42 hours 28 minutes 16 seconds. But Roemer noticed that the eclipses do not occur at exactly equal intervals of time; the intervals are somewhat *longer* than average when Jupiter is moving away from the earth, and somewhat *shorter* than average when the two planets are approaching each other. Roemer concluded, quite correctly as it turned out, that these observations must be a result of the finite velocity of light. He reasoned that if the speed of light were infinitely great, then the motions of the two planets would not enter into the matter and the intervals between eclipses would all be equal.

Here's how Jupiter's moon can be used to measure the speed of light. First of all, the time necessary for the moon to make 1 revolution must be determined. This can be done by averaging the time intervals between eclipses starting at a time when Jupiter and the earth are closest together, continuing through the time when they are farthest apart, to the time when they are back at the starting point. Having completed 1 circuit, the effect of the speed of light will have canceled out and the average period of revolution of the moon will equal the average time interval of the eclipses. Figure 26(B) shows the orbits of the earth and Jupiter to be in one plane, which is very nearly true. It also shows the locations of the planets when they are closest together (1) and far-

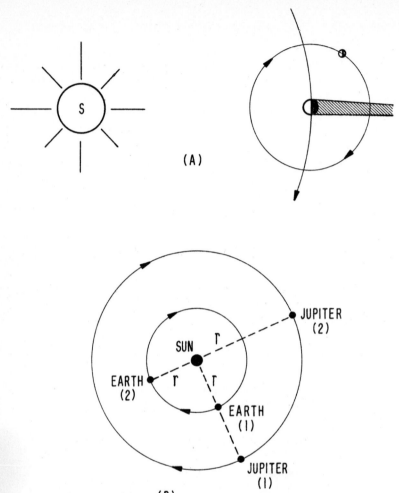

FIGURE 26. (A) One of Jupiter's moons enters the shadow of the planet once each revolution. (B) The relative positions of the earth and Jupiter when they are closest together (1) and farthest apart (2).

thest apart (2). The difference between these two distances is 2r, the diameter of the earth's orbit around the sun. This fact, and the measurements made earlier will enable us to determine the speed of light.

Referring to the geometry of Figure 26(B), you will recall that we measured the time intervals between eclipses for 1 complete circuit from minimum to maximum separation of the planets and back to minimum. This gave the true period of revolution of the moon. Let's call that time T_r. Also, let's call the apparent time interval between eclipses T_e. Then for any 1 eclipse, $T_r - T_e$ equals the difference between the actual and observed time interval of 1 revolution. While the moon is making 1 revolution, the earth and Jupiter have moved apart a certain distance, which we can call Δr.* Then the time difference $T_r - T_e$ is the time necessary for light to travel the additional distance Δr. The sum of all the time differences taken while the planets are receding from each other is the time needed for light to travel a distance equal to the difference between the maximum and minimum separation of the two planets. This distance is $2r$. This means that

$$\text{Velocity of light } c = \frac{\text{diameter of earth's orbit}}{\text{sum of time differences while planets are receding}}$$

or, in the sign language of mathematics,

$$c = \frac{2r}{\displaystyle\sum_{d=\min}^{d=\max} (T_r - T_e)}$$

The denominator is shorthand for "the sum of the differences $T_r - T_e$ from the point where d (the planetary separation) is a minimum to the point where it is a maximum."

The most accurate figure for the velocity of light obtained by this method is $c = 301{,}500$ kilometers per second. This compares favorably with the presently accepted value, $c = 299{,}792.5$ kilometers per second. (This figure is thought to be correct to within plus or minus 0.4 kilometer per second.)

* Conventionally, the Greek letter Δ (delta) preceding a quantity such as r signifies a small change in r.

In playing craps, does the thrower of the dice have an advantage?

Before analyzing the question, the uninitiated may desire a brief description of the game. Two dice are thrown. If the total turned up is 7 or 11 (a natural), the thrower wins and retains the dice for the next throw. If the total turned up is 2, 3, or 12 (craps), the thrower loses but retains the dice. Any other number that turns up becomes the thrower's *point* and he must continue to throw, in an attempt to *make his point;* that is, he must throw this number again before throwing a 7. If successful, he wins and retains the dice. But if a 7 turns up instead, he loses the play and the dice pass on to the next player.

When two dice are thrown, there are $6 \times 6 = 36$ equally likely ways in which one of the six faces of die A and one of the six faces of die B may come up. Of the 36 possibilities, the total 2 can be made in only one way, a 1 on die A and a 1 on die B. Let's write this possibility as $(1,1)$. Three, on the other hand, can be made in two ways $(1,2)$ and $(2,1)$. Similarly, the total 6 can be made in five ways, $(1,5)$, $(2,4)$, $(3,3)$, $(4,2)$, $(5,1)$. The table below gives the results of such an analysis for each of the possible totals.

Total on dice	2	3	4	5	6	7	8	9	10	11	12
Frequency of occurrence	1	2	3	4	5	6	5	4	3	2	1

This table gives all the information needed for an analysis of the game. First of all, the throw of the dice can result in any 1 of 36 combinations. The probability of throwing a particular total is found by dividing its frequency of occurrence by 36. The probability of throwing a 5, then, is $\frac{4}{36} = \frac{1}{9}$. The probability of throwing one of a number of specified totals, say of throwing a 7 or 11, is equal to the sum of their separate probabilities; in this case, $\frac{6}{36} + \frac{2}{36} = \frac{8}{36} = \frac{2}{9}$. The probability of making a point, that is, of throwing a 5 (for example) before throwing a 7 is the ratio

122

of the frequency of occurrence of the point to the sum of the frequencies of occurrence of the point and 7: for the case of making 5, $\dfrac{4}{4+6} = \dfrac{4}{10} = \dfrac{2}{5}$. Finally, the probability of throwing a particular number, and then of making it is the product of the two separate probabilities; the probability of throwing a 5, for example, and then making it is $\dfrac{4}{36}\dfrac{2}{5} = \dfrac{8}{180} = \dfrac{2}{45}$.

By using the methods outlined above, it's possible to prepare a table of probabilities of all possible results of one play.

The thrower will win	*Probability*
1. By throwing a natural (7 or 11).........	$\dfrac{6+2}{36} = \dfrac{2}{9} = \dfrac{220}{990}$
2. By throwing and making 6 or 8........	$2\dfrac{5}{36}\dfrac{5}{5+6} = \dfrac{125}{990}$
3. By throwing and making 5 or 9........	$2\dfrac{4}{36}\dfrac{4}{4+6} = \dfrac{88}{990}$
4. By throwing and making 4 or 10.......	$2\dfrac{3}{36}\dfrac{3}{3+6} = \dfrac{55}{990}$
Probability he will win	$\dfrac{488}{990} = \dfrac{244}{495}$

The thrower will lose	*Probability*
1. By throwing craps (2, 3, or 12)	$\dfrac{1+2+1}{36} = \dfrac{55}{495}$
2. By throwing and losing 4 or 10	$2\dfrac{3}{36}\dfrac{6}{3+6} = \dfrac{55}{495}$
3. By throwing and losing 5 or 9	$2\dfrac{4}{36}\dfrac{6}{4+6} = \dfrac{66}{495}$
4. By throwing and losing 6 or 8	$2\dfrac{5}{36}\dfrac{6}{5+6} = \dfrac{75}{495}$
Probability he will lose	$\dfrac{251}{495}$

As you can see, it turns out that the thrower is at a slight disadvantage since the odds against him are 251 to 244. In spite of the almost universal desire among players to roll the dice, a smart gambler will defer to his eager opponents.

What is algebra?

The word *algebra* is usually used to mean the body of rules used for solving problems about numbers with the aid of abbreviations, such as letters, and operative signs of one sort or another. The transition from the use of words to algebraic shorthand appears to have grown gradually over the period from about 1200 to 1600 A.D. The Arabs and Hindus were the first to use the square root sign $\sqrt{\ }$. The signs $+$ and $-$ were used in medieval warehouses to signify whether the weight of a shipment was above or below the expected amount. Widman's *Commercial Arithmetic,* published in 1489 in Leipzig, first put them in general use. Record's *Commercial Arithmetic,* an English publication of the following century, introduced the operators $\times$ and $=$. Other mathematicians replaced words by symbols and letters, and soon algebra came to have the form that we recognize today. The apparently sterile jumble of letters, numbers, and symbols that frightens thousands of high-school sophomores each year is nothing more than a convenient shorthand that releases mathematics from the inaccuracies and redundancies of everyday speech. When we speak of setting up an equation, what we really mean is that we are going to put a problem into a simple form in which its meaning is obvious. Algebra is nothing more than a body of rules telling us how to do this. It has a language all its own that we must learn much as we learn French, German, or Latin. Equations are nothing more than mathematical sentences. The really difficult hurdle is learning how to translate problems from English into the language of algebra.

An example may illustrate how this is done.

The problem. A train leaves New York for Philadelphia at 2 o'clock traveling at 60 miles per hour. A second train leaves Philadelphia for New York at 3 o'clock traveling at 40 miles per hour. If New York is 110 miles from Philadelphia, when do they meet?

The solution. First of all, we will want to get a plan of attack. We know the speed of each train and the time of day that it leaves, so we can write an expression for the distance that it has traveled at any time after departure. What we must find is the time that elapses before both trains are the same distance from either city. Let's write the facts that are known:

Fact 1. A train leaves New York at 2 o'clock traveling at 60 miles per hour.

$$D = 60T \tag{1}$$

That is, the distance D that the train travels from *New York* is equal to the speed, 60 miles per hour, times the elapsed time in hours T.

Fact 2. A second train leaves Philadelphia for New York at 3 o'clock traveling at 40 miles per hour.

$$d = 40(T - 1) \tag{2}$$

That is, the distance d that the train travels from *Philadelphia* is equal to the speed, 40 miles per hour, times the elapsed time in hours $(T - 1)$. We use $(T - 1)$ because the second train leaves the station 1 hour after the first train and travels a total of 1 hour less.

Fact 3. New York is 110 miles from Philadelphia.

Notice from the equations given above that:

$$D = \text{distance first train from New York}$$
$$d = \text{distance second train from Philadelphia}$$

When the trains meet, they will be at the same point. Then:

$$D + d = 110 \tag{3}$$

Or the distance of the first train from New York, plus the distance of the second train from Philadelphia must equal 110 miles. Now let's add equations (1) and (2).

$$D = 60T$$
$$\underline{d = 40(T - 1)}$$
$$D + d = 60T + 40(T - 1) \tag{4}$$

125

But equation (3) tells us that $D + d = 110$, so we can substitute this number in equation (4).

$$110 = 60T + 40(T - 1) \qquad (5)$$
$$110 = 60T + 40T - 40$$
$$110 = 100T - 40$$
$$150 = 100T$$
$$1.5 = T = \text{hours after 2 o'clock}$$

Since $T = 1.5$ hours after 2 o'clock, the trains will meet at exactly 3:30 o'clock.

If this exercise has you creaking at the mathematical seams, just imagine how difficult such a problem would have been to the most brilliant Greek mathematician. He had no way of breaking such problems down into short mathematical sentences. As a matter of fact, he had used up all the letters of the alphabet for proper numbers so there were none left for algebraic symbols. Because of the rules of algebra, we can break a complex problem down into bite-sized pieces that are easy to handle. After combining the pieces, we obtain an equation, such as equation (5), which contains all the information given in the statement of the problem. It is then only necessary to follow a few simple mechanical rules in order to obtain the answer.

What are directed numbers?

Most of us have become familiar with negative numbers in our everyday lives. We speak of $-10°F$, D day minus 1, and the like. The physical meaning of these numbers is plain to see; they represent a quantity measured from some arbitrary point in a direction that is just the opposite of the conventional or positive one. We have also learned that squaring a negative number gives a positive result: $-2 \times (-2) = +4$. Consequently, every number, whether positive or negative, is the square root of a *positive* number. Granting that fact, then what kind of a number can be the square root of a negative number? Take $\sqrt{-4}$, for example. The square root of -4 can't be -2, for that is the square root of $+4$. Then what is the square root of -4? Or -23? Or of any negative number?

126

Since $\sqrt{-4} = \sqrt{+4 \times (-1)} = 2\sqrt{-1}$, the real question is: What is $\sqrt{-1}$? I think we can illustrate this best with the aid of Figure 27. The distance OA in the diagram represents 2 units of length to the right of an arbitrary starting point O. This is the direction usually taken as positive, so $OA = +2$. OB, on the other hand, is measured to the left, or negative direction, so $OB = -2$. Here we have the point of the matter. If we take any number, such as $+2$, which is represented by OA in the diagram, and multiply by -1, we get the number -2 which is represented by OB, opposite to OA. If we again multiply by -1, OB changes

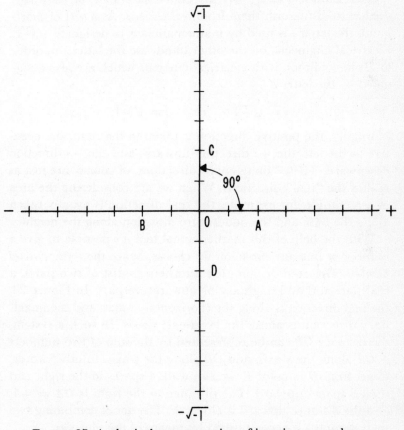

FIGURE 27. A physical representation of imaginary numbers.

127

back to *OA*. In other words, *multiplying by* − 1 *rotates a line through a half circle*, or 180°.

If we multiply $+2 \times \sqrt{-1} \times \sqrt{-1}$, we are really multiplying $+2 \times (-1)$, so the effect on the direction of the line is once again a change of 180°. If this is so, then multiplying by $\sqrt{-1}$ must rotate the line just *half* that amount, or 90°. The distance, $2\sqrt{-1}$, therefore, can be represented by *OC* in the diagram. Multiplying twice by $\sqrt{-1}$ rotates a distance through an angle of 180°. Each time we multiply a number by $\sqrt{-1}$, we rotate the number through an angle of 90°.

Such numbers as $2\sqrt{-1}$ are called *quadrature*, or *imaginary*, *numbers* to distinguish them from *real numbers*. As a sort of shorthand, the letter *i* is used by mathematicians to designate $\sqrt{-1}$. Electrical engineers, on the other hand, use the letter *j* in order to avoid confusion with electrical currents which are also designated by the letter *i*.

$$i = \sqrt{-1} \qquad \text{or} \qquad j = \sqrt{-1}$$

Normally, the positive direction is taken to the right, the negative to the left, the $+i$ direction upward, and the $-i$ direction downward. These "imaginary" directions, of course, are just as real as the "real" directions. When we are considering the area of a rectangle, for example, the real direction is usually taken along the base and the quadrature direction along the height.

With the help of this mathematical tool it's possible to give a number or line any direction we choose, hence the term, *directed numbers*. Directed or complex numbers consist of two parts, a real part and an imaginary or quadrature part. In Figure 28, the real direction is along the (horizontal) *x*-axis, and the imaginary direction is along the (vertical) *y*-axis. In such a system, the distance *OP* can be represented by the sum of two numbers —*OA* along the *x*-axis, and *OB* along the *y*-axis. In other words, to get from *O* to point *P*, we can walk 4 spaces to the right and then 3 spaces upward. The distance to the right is $OA = +4$, and the distance upward is $OB = 3i$. The act of combining two such distances is accomplished by means of the plus sign.

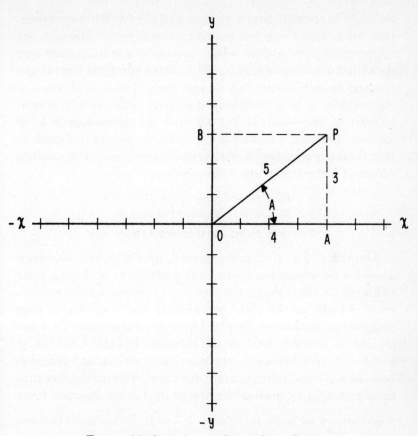

FIGURE 28. Complex or directed numbers.

$$OP = OA + iOB = 4 + 3i$$

Such numbers have enormously extended the usefulness of algebra, since they provide a numerical representation of a number having a direction. To illustrate the use of complex numbers, imagine a boat acted upon by two forces: an engine which drives it north at 3 miles per hour; and the wind which blows it in an easterly direction at 4 miles per hour. What is the true speed and direction of the boat?

The $+i$-direction will represent north, the $-i$-direction south,

129

the right horizontal direction, east, and the left horizontal direction, west. In solving this problem, our directed numbers will represent *velocity,* a term which physicists use to include both speed and direction of motion. With this in mind, the first velocity (caused by the engine) can be represented by $0 + i3$, since all the motion is in a northerly direction. The second velocity (caused by the wind) is $4 + i0$, since all the motion is in an easterly direction. (The zeros are included merely to remind us that these are numbers having specific directions.) The resulting direction is the sum of the two *components,*

$$0 + i3 = \text{engine component of velocity}$$
$$\underline{4 + i0 = \text{wind component of velocity}}$$
$$4 + i3 = \text{speed and direction of boat}$$

The velocity $4 + i3$ is the answer, although by changing it around a bit we can put it in a more useful form. Referring again to Figure 28, the complex number $4 + i3$ tells us to start out at O. move 4 units to the right (because of the $+4$), and 3 units straight up (because of the $i3$). These two components ($+4$ and $i3$) add up to the line OP on the diagram. But $(OP)^2 = 3^2 + 4^2$, so $OP = 5$. In other words, the boat *actually* moves at 5 miles per hour in a general northeasterly direction. To find the direction more precisely, we must determine angle A in the diagram. From trigonometry we know that $\tan A = \dfrac{3}{4} = 0.75$. A table of tangents gives angle $A = 36°52'$. The complete answer, then, is 5 miles per hour in a direction $36°52'$ north of east.

How many colors are needed for the most complicated map?

It's common practice in map making to color the various political subdivisions in such a way that no two adjacent regions will have the same color. In other words, no boundary may have the same color on both sides. What is the minimum number of colors that a cartographer must have at his disposal in order to meet this requirement for the most complicated map imaginable? Ten? Twenty? It's easy to show that at least four colors

130

are necessary. Four are required whenever a nation or state is surrounded by three others. To prove the point to yourself, merely draw a circle with three external lines touching the circumference. You will soon agree that four colors are necessary. But try as they may, no one has ever designed a map, whether planar or spherical, that requires more than four colors. No matter how complicated the map, four colors always seem to suffice.

The so-called *problem of the four colors* has been attacked by mathematicians for generations and yet no proof has been found. The best that has been accomplished to date is to show that not more than five colors are ever necessary. The proof is fairly complicated and so will not be presented here, but the interested reader can find it in various books on topology.

The matter can be cleared up either by constructing a map that requires five colors, or finding a proof that only four colors are ever necessary. If you decide to pursue the former course, working with a flat map can also solve the problem for a globe because, having a globe, we can always make a small hole in one of the colored regions and "open up" the resulting surface to form a plane.

Surprisingly enough, the problem of the colors, which has so successfully eluded solution for the simple plane, has been solved with ease for a pretzel and a doughnut! Seven colors, for example, are the maximum number needed to color a doughnut-shaped map and examples have been given in which the seven are actually required.

The requirement of four colors is one that practically nobody doubts—yet no one has been able to prove. Incredible? Well, why not get some paper and a pencil and give it a whirl? Perhaps you will experience that elusive moment of inspiration needed to find the proof.

What is a rhumb line?

The directions north, south, east, and west are called the *cardinal points*. As a group, they are unique among the infinitude of directions and yet each has certain individualistic attributes.

131

North and south are the directions of the earth's poles, east is the direction of rotation of the earth, and west is the direction opposite to the rotation of the earth. North and south are limited directions in that a plane flying north will eventually pass over the Pole and begin flying south. As a matter of fact, south is the only direction it can fly from the North Pole. East and west, on the other hand, are unlimited directions. A plane can take off and fly east or west indefinitely without ever changing its heading.

But there are many other directions—the *intercardinals*—which lie between the directions founded on the earth's rotation. What about these? What kind of directions are they?

Although it may seem unbelievable at first glance, all the intercardinal directions are spirals! Or more properly, *loxodromic curves*. Suppose you were to fly in a northeasterly direction indefinitely. You would go around and around the earth, always maintaining an angle of 45° with every meridian of longitude that you cross. Each time you cross the original meridian you would be closer to the Pole but, in theory at least, you would never quite reach it. Straightening out this spiral has been one of the major problems of navigators since man first learned the nature of the sphere we live on. When you travel in any direction other than the cardinals, the spiral path that you follow is called a *rhumb line*. If you steam straight on a course of 45°, your wake looks like a straight line, and your gyrocompass always points north at an angle of 45° to the ship's heading, but you are really moving on a spiral course because all the meridians converge at the North Pole. For many practical purposes, however, a rhumb line is not very different from a straight line, except near the poles, or when the distance covered is very large.

The shortest distance between two points on the globe is a great circle course, and on long voyages it is desirable to follow such a course. The following table shows the distance in nautical miles from New York City to various other cities as measured along a great circle and a rhumb line.

132

	Rhumb line	Great circle	Difference
New York to Chicago	622	619	3
New York to Paris	3,290	3,149	141
New York to Tokyo	6,932	5,856	1,076

For distances of a few hundred miles or less, the difference is slight and the rhumb line is usually chosen. The major advantage to the use of the rhumb line is the fact that it is a straight line on a Mercator chart.

How high are the lunar mountains?

To the unaided eye of the average moon-gazer, the lunar topography seems to be marked only with a few vague, dark areas, which have suggested fanciful objects to men through the ages. Some have seen the face of a man, others the profile of a woman, and still others a superbly coiffured French poodle. When Galileo looked through his telescope, however, he found the moon's surface to be covered with innumerable "seas," plains, mountains, and gorges. Although his homemade telescope had a magnifying power of only 30 diameters, it brought into view a great wealth of detail on the moon's surface. This discovery so unnerved his contemporaries that many would not even look into the telescope, for fear of disrupting their Aristotelian complacency. But Galileo persisted in his so-called madness, and noticed that some parts of the moon are very rugged, having many mountains which cast long black shadows in the light of the setting lunar sun.

The mountains of the moon run mostly in chains or groups, the highest of which rise to 25,000 feet or more above the plain. Their height is determined from the length of their shadows and a knowledge of the angle of the sun as it would appear from that point on the moon. Galileo used another method which depends upon the fact that the sun rises earlier and sets later on a high peak than on the surrounding valleys. Suppose that the moon were perfectly smooth except for a peak located as at P in Figure 29. Now visualize a ray from the sun that is tangent to the moon at T and that just manages to illuminate the top of the peak. To an observer on the earth, the peak would appear as an

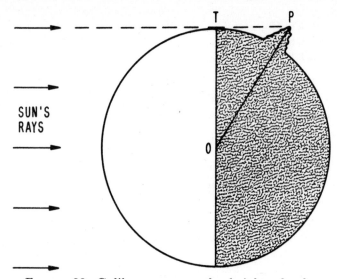

SUN'S
RAYS

FIGURE 29. Galileo measures the height of a lunar
mountain.

isolated point of light on the dark portion of the moon. The distance *TP* can then be measured in terms of the moon's apparent
diameter. This gives two sides of a right triangle which can be
solved for *OP*, which is the sum of the moon's radius (a known
quantity) and the height of the mountain.

What is the length of a degree of longitude or latitude?

A difference of 1° in latitude corresponds to the same distance
along the surface of the earth no matter where it is measured.*
This is because latitude is always measured in a north-south direction along great circles which are, by definition, the same distance in length. This can be made clear by referring to Figure 30.
Since all such circles pass through the poles, their length is equal
to the circumference of the earth. Therefore, 1° of latitude represents $\frac{1}{360}$ of the earth's circumference. If the earth's radius is taken
as 3,960 miles, this comes to

* This is only approximately true, strictly speaking, because the earth is an oblate
spheroid.

$$\frac{2\pi(3,960)}{360} = 69 \text{ miles (approximately)}$$

A difference of 1° of longitude is also equal to about 69 miles in length *when it is measured along the equator.* This is true because the equator is also a great circle and, therefore, equal in length to the earth's circumference. As we move above or below the equator, however, a degree of longitude corresponds to progressively shorter lengths. When we reach a point a few feet from the North Pole, for example, we can "walk around the world" by covering an extremely short distance. Yet this distance still corresponds to 360° of longitude.

Through the use of spherical trigonometry, it can be shown that a difference of 1° of longitude is equal to

$$69 \times \cos(\text{latitude})$$

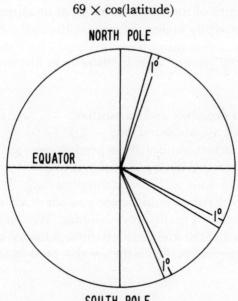

FIGURE 30. One degree of latitude corresponds to about sixty-nine miles anywhere on the surface of the earth.

135

where cos(latitude) is the cosine of the latitude of the observer in degrees. Near New York (latitude 41°), 1° of longitude would be

$$69 \times \cos 41° = 69 \times 0.7547 = 52 \text{ miles (approximately)}$$

Before we leave the subject, you may want to know the longitude and latitude of a mine dug deep in the earth. If the mine shaft is dug straight down, following a plumb line, it may go all the way to the center of the earth, and the bottom will have the same longitude and latitude as its top. For that matter, we can go in the other direction just as easily. If we were to build a monument straight up into the sky for 1,000 miles, the longitude and latitude of the top would be the same as the base. It is this technique that is used in mapping stars. When we speak of a point having a certain latitude and longitude, we happen to be earth-bound at the moment. What we are really describing is a line extending from the center of the earth to the great imaginary sphere on which the heavenly bodies seem to be attached. During a total eclipse of the sun or moon these two bodies have the same "latitude" and "longitude," and are directly in line with the center of the earth.

Why do two negatives make a positive?

Have you ever wondered why $-2 \times (-3) = +6$? Back in school, the justification usually depended upon an analogy with the well-known grammatical double negative: I have not done nothing is the same as I have done something. While this is a neat analogy designed to help young minds retain an important bit of knowledge, it really proves nothing. We can obtain a real understanding of the true situation through the use of the *distributive law of multiplication*. This is the law that permits us to write

$$2(a + b) = 2a + 2b$$

We have *distributed* the 2 over the *a* and the *b* within the parenthesis. If you are willing to accept this law, then the proof follows logically.

136

$$\begin{aligned}
(-2)(-3) &= (-2)(-3) + (0)(3) \\
&= (-2)(-3) + (-2 + 2)(3) \\
&= (-2)(-3) + (-2)(3) + (2)(3) \\
&= (-2)(-3 + 3) + (2)(3) \\
&= (-2)(0) + (2)(3) \\
&= (2)(3) \\
&= +6
\end{aligned}$$

Of course, any numbers or letters at all can be used in place of the 2's and 3's with the same result.

What is reverberation?

Reverberation is the persistence of a sound in a room for a length of time after the source has ceased emitting. It results from repeated reflections of sound waves from walls and other hard objects in the room.

In order to measure reverberation, a quantity known as *reverberation time* has been invented. It is the length of time required for a sound to reach one-millionth of its original intensity. The selection of one million for this ratio has a practical significance, since normal speech or music are about one million times as intense as a weak sound that is just audible in a quiet room. This makes possible a simple method of obtaining an approximate figure for the reverberation time of a living room, or auditorium. All that need be done is to produce a fairly loud tone and measure the duration of its audibility with a stop watch.

The time of reverberation of a room depends upon its dimensions and the degree to which its contents absorb sound. Suppose a room under test has a size such that the average distance between reflecting surfaces (the mean free path) is 20 feet. Suppose, further, that on each encounter with a wall, the sound loses 5 per cent of its intensity through absorption. After 1 reflection, the intensity drops to 95 per cent of the initial figure; after 2 reflections it falls to $0.95 \times 0.95 = (0.95)^2$; after 3 reflections, $0.95 \times 0.95 \times 0.95 = (0.95)^3$; and after n reflections it is $(0.95)^n$ times the initial intensity. Let's let n be the number of reflections necessary for the sound to fall to one-millionth of its initial intensity. Then,

$$(0.95)^n = \frac{1}{1,000,000}$$

and $$n = 224 \text{ reflections}$$

Since sound travels at about 1,120 feet per second in air, and since the average distance between reflections is 20 feet, a sound will be reflected on the average $1,120 \div 20 = 56$ times per second. The number of seconds needed for the sound to decay to one-millionth of its initial intensity is then

$$\text{Reverberation time} = \frac{\text{number of reflections}}{\text{number of reflections per second}}$$
$$= \frac{224}{56} = 4 \text{ seconds}$$

The hearing in such a room is relatively poor because of excessive reverberation. Its listening qualities can be improved, however, by adding sound-absorbing materials like carpets and draperies. Most people prefer a room having a reverberation time of about 1 second. With this fact in mind, it's possible to determine a new figure for the amount of intensity that must be lost at each reflection of the sound. For a reverberation time of 1 second in our hypothetical room, there must be precisely 56 reflections before the sound dies out; in other words, $n = 56$. If we allow L to represent the loss at each reflection, then $1 - L$ is the amount reflected at each encounter with a wall.

$$(1 - L)^{56} = \frac{1}{1,000,000}$$
$$(1 - L) = 0.81$$
$$L = 1 - 0.81$$
$$L = 0.19$$

The sound absorption of the room must be increased from 5 per cent to 19 per cent in order to achieve a time of reverberation of 1 second.

How much energy does the sun radiate?

Only a portion of the sun's radiation is given off in the form of sunlight. Much of the solar output arrives at the earth in the form of wavelengths that are either too short or too long to make any visual impression. We have all heard of infrared and ultra-violet light. These and other radiations are emitted by the sun and arrive at the earth along with the light we know so well. But all such radiations have one thing in common: they produce heat when absorbed by the earth. So the most effective way to measure the sun's total radiation is by the heat it produces.

Precise measurements of the heating effect of this radiation indicate that solar energy would arrive at a rate of 4.7 million (4.7×10^6) horsepower per square mile of the earth's surface if the atmosphere were withdrawn. This amounts to 1.5 horsepower per square yard! Since we know the rate of radiation at the earth's distance from the sun, it's a simple matter to calculate the total amount of energy being radiated by the sun. Imagine a great spherical shell that completely encloses the sun. It's clear that all the sun's radiation must fall on the surface of this shell. If we make the shell's radius equal to the earth's solar distance, 93,000,000 miles, it will have a surface area of $4\pi(93,000,000)^2$ or about 1.1×10^{17} square miles. The total amount of energy falling on the shell corresponds to $1.1 \times 10^{17} \times 4.7 \times 10^6 = 5.2 \times 10^{23}$ horsepower! This amounts to about 70,000 horsepower per square yard, at the sun's surface. Each square *centimeter* develops over 8 horsepower continuously!

The computations given above assume that the sun radiates energy at the same rate in all directions—an assumption against which there is not the slightest evidence. On the same assumption, only about one two-billionth of the solar radiation is received by the earth. A mere one hundred-millionth is intercepted by all the planets and all their satellites combined. Most of the sun's energy by far travels out into space, and to the best of our knowledge, on and on forever.

How many satellites are required for transatlantic communications?

The National Aeronautics and Space Administration of the United States has announced that a giant balloon satellite, 100 feet in diameter, will be put into orbit in order to reflect radio waves between two distant points on the earth. The satellite, which will be brighter than the North Star, will be made of light plastic covered with a thin aluminum coating and will weigh 150 pounds. A transmitting antenna at one point on the earth will send radio energy toward the satellite, which will reflect some of it back to earth in all directions. A sensitive receiving station, perhaps thousands of miles away, will be able to pick up the weak but intelligible signals.

Since the Atlantic crossing between Newfoundland and Scotland is about 2,000 miles in length, a question of interest concerns the percentage of the time that a satellite will be visible—that is, above the horizon—at both places. At other times, of course, communication between the two places will not be possible since one or the other antenna will be unable to "see" the satellite.

It can be shown that a satellite at an altitude of 1,000 miles will be "visible" at both places about 5 per cent of the time. It would appear that continuous communication can be provided merely by using a sufficient number of satellites, the number being $\frac{100}{5}$ or about 20 satellites. The difficulty with such an arrangement is the great precision required in the motions of the various satellites. Each must come into view just as its predecessor passes over the horizon. The difficulties connected with establishing such synchronized, equally spaced satellites are probably beyond the capabilities of our present technology.

With this limitation in mind, it's possible to calculate the number of random satellites that would be required for almost constant transatlantic communication.

If 1 satellite is visible, that is, above the horizon, at both places, 5 per cent of the time, it will be invisible $100 - 5 = 95$ per cent of the time. If we consider a unit period of time, the satellite will be invisible for the fractional period 0.95. If a second satellite is placed in orbit with a similar but slightly different orbit and speed, both satellites will be invisible $0.95 \times 0.95 = (0.95)^2 = 0.9$, or for a fractional period equal to 0.9. If 3 satellites are placed in orbit, the fractional time becomes $0.95 \times 0.95 \times 0.95 = (0.95)^3 =$

140

0.86. In general, if n satellites are used, the fractional period of invisibility will be $(0.95)^n$. Suppose we require a communication path to exist at least 99.9 per cent of the time. The corresponding period of invisibility is $100 - 99.9 = 0.1$ per cent, or 0.001 in fractional form. If we place this small fraction equal to $(0.95)^n$ we obtain

$$(0.95)^n = 0.001$$
$$n \times \log 0.95 = \log 0.001$$
$$n = \frac{\log 0.001}{\log 0.95}$$
$$n = 135 \text{ satellites}$$

The prospect of placing 135 man-made radio reflectors in the sky does seem staggering, but the number can be reduced considerably by using higher orbits.

What is the principle behind whispering galleries?

Look at any circle from an angle and you will see an ellipse.

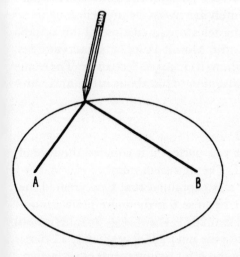

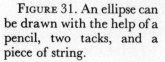

FIGURE 31. An ellipse can be drawn with the help of a pencil, two tacks, and a piece of string.

You can draw one easily with the aid of two tacks and a piece of string, as shown in Figure 31. One end of the string is attached to each tack. The pencil draws the string tight and moves all the

141

way around tracing out an ellipse. In mathematical terms, an ellipse is the path of a point in a plane such that the sum of its distances from two fixed points is constant. Since the length of the string does not change, the path of the pencil conforms with the definition.

The points at which the tacks are located, A and B, are called the foci of the ellipse. It is a characteristic of these curves that light emanating from one focus will be reflected by the ellipse and will all return to the other focus.

It is this reflecting characteristic of the ellipse that accounts for the sound effects obtained in certain buildings with arched ceilings. If the ceiling is elliptical in shape and if the two foci are accessible, it's possible to stand at one focus and hear the subdued tones of a conversation taking place at the other focus. Move away from the focus and the communication link vanishes.

The parabola is an ellipse one of whose foci has been moved infinitely far away. For this reason, a light source placed at the focus of a parabola will produce parallel rays which reach out toward the other focus infinitely far away. Many reflectors from flashlight mirrors to reflecting telescopes make use of this principle. The 200-inch mirror at the Mount Palomar Observatory is a paraboloid that reflects light to its focus 55 feet away. The same principle is used in the construction of parabolic radar and communications antennas.

What are perfect numbers?

To the Greeks, a number was perfect if it equaled the sum of all its divisors—except itself. The first such number is $6 = 1 + 2 + 3$. In ancient times, it was explained that God created the world in 6 days instead of 1 because 6 is the more perfect number. The Greeks also noticed that $28 (1 + 2 + 4 + 7 + 14)$ is also the sum of its divisors. During the next 2,000 years only 11 more numbers were found meeting the strict requirements of perfection. In addition to perfection, the Greeks attached other interesting characteristics to numbers. The number *one* was identified with reason; *two* with opinion; *four* with justice; *five* with marriage; *seven* with health; and *eight* with love and friendship. To the Greeks, the

142

number *four* was really four dots arranged in the form of a square, and since it was the product of two equals, it was associated with justice. Hence, *foursquare* and *square shooter* are still associated with a just man. Even numbers were feminine and odd numbers masculine. As a result, even numbers represented evil and odd numbers, good. The characteristic of even numbers that most irritated the Greeks was their ability to be bisected, as 16 into 8, 8 into 4, 4 into 2, and 2 into 1. This sort of thing suggested the infinite, a subject that was extremely distasteful to the Greeks. The odd numbers, on the other hand, put a stop to this sort of thing by producing improper fractions which resisted further bisection.

The number *five* represented marriage because it was the union of the first masculine and feminine numbers. The Romans had a slightly different slant on this point, however, attributing love to *six* since it was the product of the two sexes.

The Pythagoreans had a special affection for the ideal number *ten* because it contains all the geometric forms. It's the sum of 1 (the point), 2 (the line), 3 (the plane), and 4 (the cube, or solid).

Two numbers were amicable if each was the sum of the divisors of the other, such as 220 (1, 2, 4, 5, 10, 11, 20, 22, 44, 55, and 110), and 284 (1, 2, 4, 71, and 142).

As you can see, the Greeks were obsessed with numbers for their own sake. This caused them to build up a body of pseudo-scientific knowledge based, to a large extent, on interesting but irrelevant ideas about numbers. Unfortunately, the religious sophists of medieval Europe seized upon this natural philosophy and gave it the powerful backing of the Church. Eventually, men like Copernicus, Kepler, Galileo, and Newton changed this philosophy (at great personal sacrifice) from one of attempting to bend nature to the preconceived notions of men, to the doctrine that nature must be studied quantitatively, and her secrets discovered. Numbers still form the basis of this philosophy, but now they are the numbers of nature rather than of men.

Are there any short cuts in multiplication?

Some short cuts in multiplication are fairly obvious and many of us have rediscovered them by accident in dealing with numbers. To multiply by 25, for example, merely multiply by 100 and divide by 4:

$$52 \times 25 = \frac{52 \times 100}{4} = 1,300$$

Multiplying by 26 or any other number close to 25 is almost as easy:

$$52 \times 26 = 52 \times 25 + 52 = 1,300 + 52 = 1,352$$

In much the same way, multiplying by 12.5, 50, 125 and 250 can be reduced to simple division by 8, 4, or 2.

Other short cuts are somewhat more sophisticated and depend upon the nature of our decimal system. Any number having two digits can be represented in the form $(10a + b)$ where a and b are any digits from 0 to 9. If $a = 3$, and $b = 4$, then $10a + b = 34$. The multiplication of two such numbers would then be as follows:

$$(10a + b)(10c + d) = 100ac + 10ad + 10bc + bd$$
$$= 100ac + 10(ad + bc) + bd$$

This might be called a general formula from which many rules of multiplication can be deduced. Suppose that we want to develop a rule for squaring a number that ends in 5. Since we are dealing with one number rather than two, $a = c$ and $b = d$. And since the number ends in 5, $b = d = 5$. Substituting this information in the formula, we obtain

$$(10a + b)(10c + d) = 100\,ac + 10(ad + bc) + bd$$
$$= 100a^2 + 10(5a + 5a) + 25 \text{ (since } a = c \text{ and}$$
$$b = d = 5)$$
$$= 100a^2 + 100a + 25$$
$$= 100(a^2 + a) + 25$$
$$= 100a(a + 1) + 25$$

This result means that the square will have 5 in the units place, a 2 in the tens place, and $a(a + 1)$ in the hundreds place. The rule is then as follows: To square a number ending in 5 (such as 35), write down 25 as the last two digits of the product; then

144

multiply the number in the tens column (3 in our example) by that number increased by 1 (3 + 1 = 4), to obtain the first two digits of the answer.

$$35 \times 35 =$$
$$5 \times 5 \; = \quad 25$$
$$3 \times 4 \; = 12$$
$$\overline{1225} \quad \textit{Ans.}$$

You can follow the procedure given above and develop many of your own short cuts in multiplication. Perhaps you will want to verify the rule for multiplication by the teens: $18 \times 15 = ?$

a. To the first number add the units digit of the other and annex a zero.

b. To this result add the product of the units digits of the numbers.

$$18 + 5 \; = 23$$
$$23 + 10 = 230$$
$$5 \times 8 \; = \quad 40$$
$$\overline{270} \quad \textit{Ans.}$$

Before we leave the subject, there are two other techniques of multiplication that ought to be mentioned. These are the methods of multiplying by complements and supplements.

The *complement* of a number is the figure required to bring it up to 100. So 5 is the complement of 95, and 7 is the complement of 93.

$$95 \times 93 = ?$$

a. Multiply the complements of the numbers and write the product as the last two digits of the answer ($5 \times 7 = 35$).

b. Subtract either complement from the other number (say, $93 - 5 = 88$) and write the difference as the first two digits of the answer.

$$7 \times 5 = \quad 35$$
$$93 - 5 = 88$$
$$\overline{8835} \quad \textit{Ans.}$$
$$145$$

The *supplement* of a number is the amount by which it exceeds 100. The supplements of 105 and 106 are 5 and 6 respectively.

$$105 \times 106 = \ ?$$

a. Multiply the supplements and write the product as the last two digits of the answer ($5 \times 6 = 30$).

b. Add the supplement of either number to the other number ($106 + 5 = 111$) and write down the sum 111 as the first figures of the answer.

$$
\begin{array}{r}
6 \times 5 = \quad\ 30 \\
106 + 5 = \underline{111} \\
11130 \quad \textit{Ans.}
\end{array}
$$

Do any rivers flow uphill?

The oblate spheroid on which we live has an equatorial diameter that is about 27 miles greater than its polar diameter. This means that a point on the equator is about 13 miles or so farther from the earth's center than are the poles. Since the directions "up" and "down" refer to one's distance from the center of the earth, the equator is some 13 miles "higher" than the poles. In this sense, a person walking toward either pole is going "downhill." Conversely, rivers that flow toward the equator are actually flowing from a lower point to a higher one—or uphill! Once we are satisfied that rivers can flow uphill, we will want to know how this can come about. After all, everyone knows that water seeks the lowest possible level. Why do some rivers defy this basic principle?

The answer lies in the centrifugal force generated by the rotation of the earth. To understand how this force can move rivers uphill, let's imagine the earth to be a perfect sphere, as in the accompanying diagram (Figure 32). Now let's place a round ball at point *A* on the rotating sphere and analyze the forces acting on the ball. First of all there will be gravity which attracts the ball *toward the center of the earth*. But a second force exists as a result of the rotation of the earth and we call it centrifugal force. This force acts in a direction *at right angles to the axis of ro-*

146

tation of the earth. That is the important point. The two forces are not in line with each other. As you can see from the diagram, the combination of the two forces tends to slide the ball toward the equator. The 13-mile bulge at the equator is just sufficient to counteract this net effect of the two forces.* If the earth were

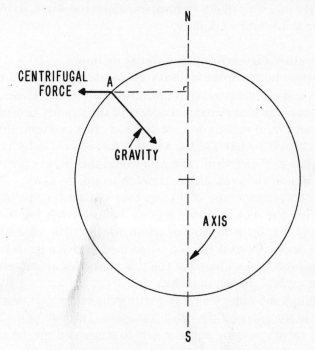

FIGURE 32. Some rivers can flow uphill with the help of centrifugal force.

to stop rotating and all other factors were to remain unchanged, portions of some rivers would reverse their direction of flow because centrifugal forces would disappear and could not assist in their uphill flow.

While part of the centrifugal force helps slide objects toward the equator, the other component tends to lift them so that their

* Of course, under these ideal conditions a river would not flow. Depending upon any slight irregularities of the terrain, a river will flow toward or away from the equator. Those flowing in the former direction are actually flowing uphill.

147

weight is diminished. A man weighing 190 pounds at the North Pole where the centrifugal force is zero would weigh 1 pound less at the equator where it is a maximum. Carrying this to extremes, if the speed of rotation of the earth should increase about 16 times, an object at the equator would weigh nothing at all! Luckily, this isn't likely to happen, since the earth is, in fact, slowing down ever so slightly.

What causes the earth's precessional motion?

During our top-spinning days we would try to spin a top so that its axis of rotation was in the vertical direction. This accomplishment was considered a true test of the master, because the top would then maintain a steady and erect position. But human skill being what it is, the axis was more likely to be inclined somewhat to the vertical. The axis would then slowly precess, or move around the vertical line through its tip.

This *precessional motion* of the top comes about because of gravity. With the axis vertical, gravity balances out because the weight of the top is distributed symmetrically with respect to the pull of gravity. But if the top is inclined, gravity tends to pull the top over. The spinning of the top resists this action, producing the conical motion of its axis.

In much the same way, the earth spins on an axis that is inclined to the plane of the moon's motion. The pull of the moon on the earth's bulging equator acts to straighten up the earth's axis by attempting to bring the bulge into its own plane of motion. But the earth's rotation tends to resist this attraction and its axis moves slowly around, much as that of the top mentioned above. The result is the precessional motion of the earth. A single turn is made in about 26,000 years at the present rate. Though it is set in its ways, even the great earth can be persuaded to move.

How can a watch be used as a compass?

The practiced woodsman can glance at the sun and tell intuitively the direction in which he should walk. You can do just

148

as well if you happen to have your watch with you. If the sun is visible, hold your watch in a horizontal plane and point the hour hand in the direction of the sun. The south direction will then be midway between the hour hand and the 12 o'clock mark on the face of the watch. If it's 4 o'clock, for example, the south direction will extend from the center of the watch toward the 2. Directions determined in this way will be approximate, of course, but useful nevertheless.

The angle between 12 and 4 o'clock in the example given above is called the hour angle of the watch. It's necessary to divide this angle in half because the hour hand moves twice as fast as the sun. It takes the sun approximately 12 hours to move across the daytime sky—an angular excursion of about 180°. But a wrist watch moves through 360° in this length of time. As you can see, the hour angle of the watch is always twice as great as the hour angle of the sun. If our clocks and watches covered 24 hours in 1 revolution of the hour hand, determination of the south direction would be much simpler. It would then be necessary merely to point the hour hand at the sun and find south always in the direction of the 24.

What are imaginary numbers?

First of all, if we are going to have imaginary numbers, we should also have real ones. *Real* numbers are the kind we are all used to, the numbers of everyday life. They can be large or small —positive or negative. We know that the square root of 4 is equal to plus or minus 2. Written in mathematical shorthand,

$$\sqrt{4} = \pm 2$$

So we can say that $+2$ and -2 are the square roots of 4. The fact that a minus times a minus is a plus means that $-2 \times (-2) = 4$, giving us the second *root* that we don't usually think too much about. But now the question comes up: Do negative numbers have square roots? Is there any such number as $\sqrt{-4}$, or $\sqrt{-12}$, or $\sqrt{-213}$? In other words, is there any number which when squared gives a negative number? The answer is yes, but

149

these numbers are not real in the sense that 2, 4, and 6 are real. We call them *imaginary numbers* and use a small letter i to indicate that they are imaginary. Thus i is defined as the square root of minus one, $i = \sqrt{-1}$.

Now let's see what happens when we multiply i by itself a few times:

$$i = \sqrt{-1}$$
$$i^2 = -1$$
$$i^3 = -\sqrt{-1}$$
$$i^4 = +1$$
$$i^5 = \sqrt{-1}$$

All even powers of i are either plus or minus 1, and all odd powers of i are plus or minus i. While i itself is not a real number, i^2 is very real indeed and equal to -1. Whenever you see $5i^2$ you will know that it is really -5, and when you see $178i^4$ you will know that you really have $+178$, and so on. While the imaginary i is really not a real number at all, it changes into a real number any time you raise it to an even power.

One of the most fascinating outcomes of all this is the fact that every number that you can imagine has exactly n n^{th} roots. In other words, pick any number, say 16: 16 has exactly 2 square roots, 3 cube roots, 4 fourth roots, 5 fifth roots, and so on. Any number at all has just n n^{th} roots. But no matter how many roots a number has, there will be no more than 2 *real* roots. The others will all contain the letter i. Take 8, for example. The cube root that we are all familiar with is 2. But there are 2 *complex* cube roots of 8. A complex number is one that consists of a real part and an imaginary part. The 2 complex cube roots of 8 are $-1 + i\sqrt{3}$ and $-1 - i\sqrt{3}$. To prove our point, let's cube one of these roots and see what we get. Remember that $i^2 = -1$ and $3i^2 = -3$.

$$
\begin{array}{r}
-1 + i\sqrt{3} \\
-1 + i\sqrt{3} \\
\hline
+1 - i\sqrt{3} \\
- i\sqrt{3} - 3 \\
\hline
+1 - 2i\sqrt{3} - 3 = -2 - 2i\sqrt{3}
\end{array}
$$

Once more

$$-2-2i\sqrt{3}$$
$$-1+ i\sqrt{3}$$
$$\overline{+2+2i\sqrt{3}}$$
$$\quad\quad -2i\sqrt{3} - (-6)$$
$$\overline{+2 \quad\quad\quad +6 \quad = 8}$$

As you can see, our complex number really does produce 8 when it is cubed. In the same way, the other root can be shown to be a cube root of 8. So 8 has 3 cube roots, 1 real, and 2 complex.

If you're wondering what value, if any, these imaginary numbers may have, you can put your mind at ease, for they are very important indeed. Electrical engineers have been using imaginary numbers for years to solve very real problems in electricity.

How far is it to the horizon?

You may have wondered, at one time or another, how far you can see on the surface of the earth from an elevated position such as a tall building or cliff. With the aid of Figure 33 we can derive an easy-to-remember formula that will give you the answer the next time the situation arises.

Let O be the center of the earth in the sketch and let h be the height of the observer above the surface. The radius of the earth, therefore, is r, and PA, the distance to the horizon, is tangent to the earth's surface at point A, which represents the edge of the horizon. Since POA is a right triangle, we know that the square of the hypotenuse equals the sum of the squares of the other two sides.

$$(PO)^2 = (PA)^2 + (OA)^2 \tag{1}$$

But distance PO is equal to $r + h$, and distance OA is equal to r. Substituting these values in equation (1) we get

$$(r - h)^2 = (PA)^2 + r^2$$

Since we are looking for PA we will rearrange the equation as follows:

151

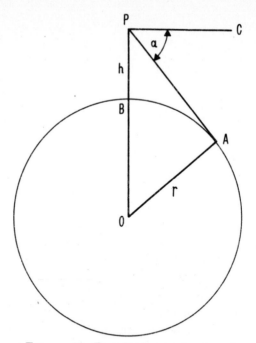

FIGURE 33. Developing a simple rule for determining the distance to the horizon.

$$(PA)^2 = (r + h)^2 - r^2$$
$$(PA)^2 = r^2 + 2rh + h^2 - r^2$$
$$(PA)^2 = 2rh + h^2$$
$$PA = \sqrt{2rh + h^2} \qquad\qquad (2)$$

A little reflection will show that h^2 must be extremely small compared to $2rh$. In one case h is multiplied by itself, a relatively small number, and in the other by twice the radius of the earth. So, as long as h is very small compared to r, we can write equation (2) as

$$PA = \sqrt{2rh} \text{ approximately} \qquad\qquad (3)$$

Now that we have the formula, let's proceed to make it a little easier to use. In order to get the correct answer using equation (3), PA, r, and h must all be in the same units, miles for exam-

152

ple. Let's change the equation so that we can substitute h in feet and get the answer PA in miles. Also, let $r = 3{,}960$ miles. Then

$$PA = \sqrt{2 \times 3{,}960 \times \dfrac{h}{5{,}280}}$$
$$PA = \sqrt{1.5h} \quad \text{miles}$$

where h is expressed in feet and PA in miles.

So, to calculate the distance to the horizon in miles, merely find the square root of $1\frac{1}{2}$ times your elevation above the earth's surface, in feet. If you happen to be located 150 feet above sea level, the horizon is just 15 miles away.

What is a cryptarithm?

One of the favorite pastimes of cryptographers is the substitution of figures in place of letters. "By way of reprisal," wrote Minos (in the journal *Sphinx* of May, 1931), "we have put letters in place of figures." A *cryptarithm*, then, is an arithmetical operation—such as addition or multiplication—in which the digits have been replaced by letters or other symbols. The reader is then challenged to find the original numbers.

Here is a sample given by M. Pigeolet, a director of *Sphinx*.

$$
\begin{array}{r}
A\ B\ C \\
A\ B\ C \\
\hline
D\ E\ F\ C \\
C\ E\ B\ H \\
E\ K\ K\ H \\
\hline
E\ A\ G\ F\ F\ C \\
\end{array}
$$

Solution. We must find the number ABC which is squared in the operation given above. Probably the first point of significance is the repetition of C as the last digit in the number and in its square. Only three digits behave in this way: 0, 5, and 6. But it can't be zero, since the first product ($DEFC$) would then be absent. So $C = 5$, or 6. Next, notice that the sum $F + H = F$. For this to be true, H must be equal to zero. This fact will help us fix the true number for C. Notice from the problem that

153

the two products $B \times C$ and $A \times C$ both end in H, which we have found to be zero. Only 5 can have more than one product ending in zero (i.e., 5 $\times$ any even digit). Six, on the other hand, has only one such product—6 $\times$ 5 = 30. So $C = 5$.

A little reflection will then show that F, the next to the last digit in the answer, must equal 2. This is true* since the square of a number ending in 5 must end in 25. If we now substitute the known numbers in the cryptarithm, the problem takes this form:

$$
\begin{array}{r}
A\,B\,5 \\
A\,B\,5 \\
\hline
D\,E\,2\,5 \\
5\,E\,B\,0 \\
E\,K\,K\,0 \\
\hline
E\,A\,G\,2\,2\,5
\end{array}
$$

Let's now investigate the middle product $AB5 \times B$.

$$
\begin{array}{r}
AB5 \\
B \\
\hline
5EB0
\end{array}
$$

We know that $B \times 5$ ends in zero; so B is an even number. Also, the digit "carried over" from $B \times 5$, when added to $B \times B$, gives a result that ends in B. In other words, $B^2 + $ (carried-over digit) $= \,?B$. Let's try substituting the even digits for B.

25	45	65	85
2	4	6	8
50	180	390	680

As you can see, 8 is the only even digit that satisfies all the requirements; so $B = 8$.

---

* The last two digits of a square are determined only by the last two digits of the corresponding number. Let the latter be a and 5. Then the units digit of the number is 5 and the tens digit is a. The right-hand part of the number, then, can be represented by $10a + 5$. Squaring this gives $100a^2 + 100a + 25$ which can be factored to give $a(a + 1)100 + 25$. It can be seen from the last expression that the product $a(a + 1)$ can have an effect on only the hundreds and thousands place of the square since the product is always multiplied by 100. If this is true, then the square must always end in 25.

154

$$\begin{array}{r} A85 \\ 8 \\ \hline 5E80 \end{array}$$

Substitution then shows that $A = 6$, the only digit that gives a product in the 5,000's. The required number is 685.

Here is another cryptarithm that you may want to solve.

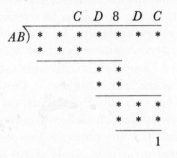

Answer. 1,089,709 ÷ 12 = 90,809.

How long does twilight last?

Twilight is the light from the sky between sunset and the full of night, or between nighttime and sunrise. It results from the scattering of sunlight that shines on the atmosphere above us. Civil twilight ends when the sun's center has sunk 6° below the horizon, because artificial illumination then becomes necessary for outdoor operations that require good light. Nautical twilight lasts until the sun's center has fallen 12° below the horizon, for at that time the sea horizon has become too dim for the navigator's sextant sights. Astronomical twilight comes to an end when the sun's center is 18° below the horizon. The fainter stars have then become visible overhead.

Because the earth's axis is tilted $23\frac{1}{2}°$ with respect to its orbit, the duration of twilight depends upon the latitude and the season of the year. It is shortest at the equator where the sun descends vertically because at that latitude it reaches the limiting angle below the horizon in the shortest time. Astronomical twilight lasts about an hour at the equator and about an hour and a half at the lati-

155

tude of New York. On June 22, it doesn't end at all north of latitude 48° north. On that date, civil twilight lasts throughout the night from latitude 60° north to 66° north. At higher latitudes the midnight sun can be seen.

It's a curious fact of nature that high-frequency radio waves also exhibit an ability to produce a twilight zone of their own. Although Marconi had shown, in 1932, that very-high-frequency radio waves (such as are now used in television broadcasting) could be sent over great distances, his discovery was ignored for over twenty years. Accepted theory said that such waves should be limited to so-called "line of sight" distances. Beyond the line of sight (that is, below the "radio horizon"), they were supposed to attenuate or fade out very rapidly. But just past the half-century mark scientists found that such radio waves produce a useful twilight zone that extends many times the line of sight distance. Although the physical principles involved are now the subject of great controversy, some theorists believe that this phenomenon, known as *tropospheric propagation*, is similar to the twilight of optical waves.

What is the reason for the lag of the seasons?

Summer in the northern latitudes begins officially on June 22, when the sun reaches the summer solstice. On that day the sun climbs highest in the sky and delivers the greatest amount of heat to the Northern Hemisphere. From June 22 on, the days become shorter and the sun moves back toward the south. Yet the hottest part of summer doesn't usually occur until August. Why does the peak of the summer heat come so long after the day of the solstice?

The earth is perhaps four billion years old and after such a great length of time its average temperature has become relatively stable. In the course of a year the earth loses about as much heat to outer space as it receives from the sun. If this weren't so, the earth's climate would change much more radically from year to year than it does. But the sun distributes its heat over the earth unevenly during the course of the year. On June 22, the northern latitudes receive a relatively high amount of heat energy, while on

156

December 22 the southern latitudes have their turn. After June 22, the sun's rays deliver less heat to the Northern Hemisphere from day to day. But until early August, its diminishing receipts are still greater than the amount of heat it loses by radiation into space. Summer does not reach its peak until the daily incoming heat has been reduced to the daily amount being lost. From that day on, the hemisphere loses more heat than it gains in a day and the weather gets progressively colder. For the same reason, the winter weather is usually the coldest in February, although the sun has turned back toward the north on December 22. The climate continues to get colder until the amount of heat received per day equals the amount lost.

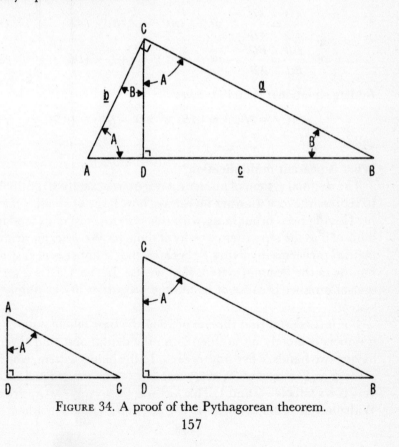

FIGURE 34. A proof of the Pythagorean theorem.

157

What is a proof of the Pythagorean theorem?

The theorem of Pythagoras states that the sum of the squares on the two sides of a right triangle equals the square on the hypotenuse. We can demonstrate a simple proof of this theorem with the help of Figure 34. Triangle ABC is a right-angled triangle with the right angle at C. Drawing a perpendicular from point C to D divides this triangle into two similar, smaller triangles ACD and BCD. Both triangles have one angle of $90°$. In addition, angles DAC and DCB are equal since their sides are mutually perpendicular. Since the angles of any plane triangle add up to $180°$, the third angles are equal and the triangles are similar.

If this is true, then

$$\frac{AD}{AC} = \frac{AC}{AB} \quad \text{or} \quad (AC)^2 = (AB) \times (AD) \tag{1}$$

$$\frac{DB}{BC} = \frac{BC}{AB} \quad \text{or} \quad (BC)^2 = (AB) \times (DB) \tag{2}$$

Adding equations (1) and (2) gives

$$(AC)^2 + (BC)^2 = (AB) \times (AD + DB) = (AB)^2$$
$$a^2 + b^2 = c^2$$

What is peasant multiplication?

The decimal system of numeration requires exactly 10 symbols to represent every number no matter how large or small it may be. Having been brought up with this system, most of us tend to think of it as the *only* system worthy of thought. We welcome arithmetical problems involving 10 because that number is so easy to handle in the decimal system that we use. In short, 10 is a very special number because of its unique properties in our number system.

But it turns out that the use of 10 as the base of our system of numbers is merely an accident—an incidental outcome of our having two hands of five fingers each. In the binary system, which is based on 2 instead of 10, every number can be represented with only two symbols—0 and 1. This fact was discovered by Gottfried Wilhelm von Leibnitz (1646–1716), one of the greatest mathema-

158

ticians that ever lived. And while they didn't realize it, the peasants of his day made use of the binary system in a simple method of multiplication called peasant multiplication. Suppose you had only learned the multiplication table through the 2's. How would you multiply 43 by 87? Give up? Well, here's how they did it in Leibnitz' day. Place 43 and 87 at the top of two columns that you are about to form. Divide 43 by 2 and divide that number by 2 and so on until you reach 1. Then multiply 87 by 2 successively as shown below.

43	87
21	174
10	348
5	696
2	1,392
1	2,784

Of course, 43 cannot be divided evenly by 2, and since we are not very good at arithmetic anyway, we throw away all extra 1's that are left over. The next step is to cross out every number in the right-hand column that is opposite an *even* number in the left-hand column. Then we merely add up the remaining numbers in the right-hand column to find the answer.

43	87
21	174
10	~~348~~
5	696
2	~~1,392~~
1	2,784

$$3,741 = 43 \times 87$$

If you will multiply 43×87 in the usual way, you will find the answer to be quite correct. And if you feel that this pair of numbers enjoys some sort of magical charm (as the Greeks would say), try the method on any other pair and you will find that it works every time—in spite of throwing away all the extra 1's.

This method of multiplication can be explained nicely in terms of the binary system of numeration. The number 43 is really

$$4 \times 10^1 + 3 \times 10^0$$

where $10^1 = 10$ to the first power, or 10, and $10^0 = 1$. Expressed in binary form, 43 is 101011, which is really

$$1 \times 2^5 + (0) \times 2^4 + 1 \times 2^3 + (0) \times 2^2 + 1 \times 2^1 + 1 \times 2^0 \quad (1)$$
$$\text{or} \quad 32 \quad + \quad 0 \quad + \quad 8 \quad + \quad 0 \quad + \quad 2 \quad + 1 = 43$$

In this representation, the symbol in each "place" of the number represents a power of 2 rather than a power of 10. In order to change a number from decimal to binary form, it is only necessary to divide it by 2 successively, as we did with 43 above, and put a 1 opposite all odd results and a 0 opposite all even results.

43	1
21	1
10	0
5	1
2	0
1	1

The binary number, 101011, is then read from the bottom toward the top. Now let's get back to peasant multiplication. To multiply 43 (which is decimal for 101011) by 87, we can multiply each term of (1) above by 87 and add up the products.

$$
\begin{aligned}
1 \times 2^0 \times 87 &= 1 \times 87 = 87 \\
1 \times 2^1 \times 87 &= 1 \times 174 = 174 \\
0 \times 2^2 \times 87 &= 0 \times 348 = 0 \\
1 \times 2^3 \times 87 &= 1 \times 696 = 696 \\
0 \times 2^4 \times 87 &= 0 \times 1{,}392 = 0 \\
1 \times 2^5 \times 87 &= 1 \times 2{,}784 = 2{,}784 \\
\hline
& \phantom{1 \times 2^5 \times 87 = 1 \times 2,784 = {}} 3{,}741
\end{aligned}
$$

Notice that what we have done is to convert 43 into binary form and multiply that number by 87.

Are there rules for the solution of figure-tracing puzzles?

The city of Königsberg and the island of Kneiphof became the center of a mathematical discussion early in the eighteenth century. Located near the mouth of the Pregel River, the area included seven bridges, as shown in the accompanying diagram. The discussion concerned whether or not it was possible to cross all seven bridges in turn without crossing any bridge a second time. Euler's famous paper *Solutio problematis ad geometriam situs pertinentis* was presented to the Academy of Sciences at St. Petersburg in 1736 in answer to this question.

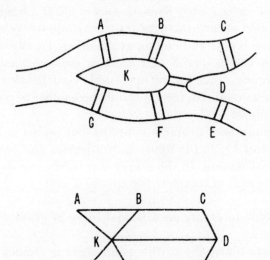

FIGURE 35. The most important figure-tracing puzzle of them all.

If we allow the islands to shrink to points, the problem can be put in a more convenient form. Referring to Figure 35, we are required to trace the given figure without retracing any part and without lifting the pencil from the paper. If this requirement is impossible, we are requested to determine the *minimum number* of strokes needed.

Before we get to Euler's conclusions, we will have to define a concept or two. First of all, we are dealing with *closed figures*—

161

figures having no free point or loose end. In any such figure, each point will have two or more lines proceeding from it. The number of such lines can be called the *order* of that point. Broadly speaking, every point can be classified into an even or an odd order. Point *B*, for example, involves three lines so its order is odd.

Euler points out that *in a closed figure the number of points of odd order is even.* This is true whether or not the figure is *unicursal* (traceable in one stroke). Also, *a figure having only points of even order can be traced by one stroke starting from any point at all.* If a figure has two points of odd order, it can be traced by one stroke starting at one of those points and ending at the other. Finally, a figure of more than two points of odd order is *multicursal* and requires more than one stroke for the operation. To determine the number of strokes required, merely count the number of points of *odd order* and divide by 2.

The Königsberg problem contains four points of odd order (*K, B, D* and *F*); so the figure is multicursal and requires two strokes in its tracing. In the practical problem, one bridge must be crossed twice to complete the circuit.

Is it possible to create an artificial force of gravity for space travelers?

The space above the earth's atmosphere is rapidly becoming littered with assorted Sputniks, Explorers, Vanguards, Discoverers, and the like. Sooner or later someone is going to shoot off a really large one carrying a human cargo and man will have truly entered the space age. At first these manned satellites will be small and crude affairs with little thought to the fine points of gracious living. But as time goes on, the manned satellite will give way to the more sophisticated space stations which have been designed and described with increasing frequency over the past decade or so.

A major inconvenience associated with these space structures, however, is the lack of a force of gravity of any consequence. An object orbiting around the earth may be thought of as a stone

162

whirling around at the end of a piece of string. The whirling motion of the satellite produces a centrifugal force which just balances the force of gravity, producing a weightless condition for objects (and persons) on the satellite. We are all familiar, of course, with the picture of a space man floating around within his space ship, surrounded by a sea of floating objects of all descriptions. But other problems are even more disconcerting. Water, for example, will refuse to pour from pitcher to glass. Worse still, once negotiated into the mouth it will be swallowed only with great difficulty since gravity is not there to do its customary part.

To solve these problems, most space-station designs (such as an early one by Dr. Wernher von Braun) resemble a bicycle wheel rotating about its central axis. This rotary motion would produce a centrifugal force which would cause all objects to fall from the center toward the rim. It would be impossible to distinguish this phenomenon from true gravity. The center of the wheel would be "up" while the rim would be "down." The portion of the space station analogous to the bicycle tire would be the habitable portion filled with oxygen, food, equipment, and other niceties of space-station life. Space passengers waiting for the next flight to Mars would stroll along the revolving "tire" completely at ease and unaware that 5 or 10 revolutions per minute of their space station makes the difference between comfort and chaos. To produce the same effect on a trip to Mars, the space ship could be made to spin on its axis. In that case, "up" would be toward the central axis, while "down" would be toward the outside shell of the ship. It might be possible to reach "up" and borrow a magazine from another passenger sitting on the opposite side of the space ship!

All this is possible because motion in a circle produces an acceleration that can take the place of the acceleration due to gravity. If you are wondering how acceleration came into the picture, perhaps we had better come back to earth for a moment. When an object is dropped near the surface of the earth, it does more

163

than fall—it *accelerates.* That is, during each succeeding instant it travels at a higher velocity than during the preceding one. It is said to accelerate at a rate of 32.2 feet per second per second. It may help to think of acceleration as the "pick-up" in a good automobile.

Suppose a car starts from rest and reaches a speed of 4 feet per second at the end of 1 second. By the end of 2 seconds it is traveling at 8 feet per second and so on, increasing its speed by 4 feet per second during each second of travel. So long as the car keeps this up it is said to have a pick-up, or acceleration, of 4 feet per second *each second,* or, more properly, 4 feet per second per second. Similarly, an object falling under the influence of gravity has an acceleration of 32.2 feet per second per second. In order to duplicate this acceleration, a space vehicle must spin it at a rate*

$$T = \frac{1}{2\pi}\sqrt{\frac{g}{R}} \quad \text{revolutions per second}$$

where $g = 32.2$ feet per second per second and R is the distance from the point of interest to the center of rotation. It may be worth noting that this artificial gravity can be tailored to any desired figure merely by spinning the vehicle at the corresponding rate.

What is Bode's law?

Bode's law is a convenient way of remembering the approximate distances of the planets from the sun. Johann Elert Bode, a Berlin professor of astronomy, pointed out an amazing series of numbers that closely represents the relative planetary distances from the sun. The series is as follows:

*The centrifugal acceleration $\frac{V^2}{R}$ is placed equal to g and the resulting equation is solved for $V = \sqrt{gR}$. But the angular velocity, $T = \frac{V}{2\pi R}$, or $V = 2\pi RT$. Substituting V in the first equation yields $2\pi RT = \sqrt{gR}$, or $T = \frac{1}{2\pi}\sqrt{\frac{g}{R}}$.

```
Mercury ........ 4 = 4
Venus .......... 7 = 4 + 3
Earth .......... 10 = 4 + (3 × 2)
Mars........... 16 = 4 + (3 × 2 × 2)
? ............. 28 = 4 + (3 × 2 × 2 × 2)
Jupiter ........ 52 = 4 + (3 × 2 × 2 × 2 × 2)
Saturn ........ 100 = 4 + (3 × 2 × 2 × 2 × 2 × 2)
Uranus ........ 196 = 4 + (3 × 2 × 2 × 2 × 2 × 2 × 2)
```

The number for each term of the series represents 10 times the planet's approximate distance from the sun in astronomical units. (An astronomical unit is the earth's mean distance from the sun, or one-half the major axis of the earth's orbit. It's usually taken to be 93,000,000 miles.) To illustrate, Bode's law tells us that Mars is $16 \div 10 = 1.6$ astronomical units from the sun. The true distance is 1.524 astronomical units, or about 4.9 per cent less. All the other distances given above are somewhat closer to the correct figure. The earth's distance must be correct, of course, because of the definition of the astronomical unit.

As astronomers considered Bode's series, it took little imagination to postulate the existence of an undiscovered planet corresponding to the fifth term of Bode's series between Mars and Jupiter. The discovery of Uranus in 1781 by William Herschel had fit so nicely into Bode's series (19.191 astronomical units), that the gap between Mars and Jupiter became all the more disturbing. Toward the end of the eighteenth century a number of astronomers organized a planned search for the illusive fifth planet.

While this was going on, a young philosopher by the name of Georg Wilhelm Friedrich Hegel was thinking along different lines. Still subscribing to the ancient and mystical ideas of numerology, he believed that there could be only seven planets, the earth and moon included. Since the days of the Babylonians, the seven wandering heavenly bodies had been the basis for the nonsense that mystics practiced with "sacred" numbers. In view of the discovery of Uranus, it took great obstinacy to maintain such a position. But this young Hegel did. He prepared a paper

165

using the strictest logic to prove that there could not be more than seven planets and so it would be quite impossible for any more to be found.

The paper came off the press just as news arrived from Palermo that a Sicilian monk had discovered a new kind of heavenly body within the solar system—the tiny planet Ceres. Though not one of the organized searchers, Giuseppi Piazzi found the small planet on the first of January, 1801, and made a few observations of its position. Unfortunately, after observing it for a short time, Piazzi became ill. Before he recovered, and before news of the discovery reached other astronomers, the tiny planet had moved to a position unfavorable for rediscovery. It was in danger of being lost in the multitude of stars. Calculating an elliptical orbit for it from the few figures that Piazzi had been able to garner was an extremely difficult, if not impossible, task. Fortunately, a young mathematician, Karl Friedrich Gauss, had just invented his method of least squares which permitted him to perform the task from only three observations. From the calculated orbit of Ceres, astronomers located the little planet on the last day of the year 1801. But of even greater interest, Ceres' orbit fulfilled Bode's exacting demands. The only flaw, of course, was Ceres' diminutive size—merely 488 miles in diameter. Such a small piece of rock could hardly be considered a planet.

If Bode's law were to be meaningful, there must be many more such tiny planets in the fifth term of Bode's series. Sure enough, these were found. The next year Pallas (304 miles) was found, followed by Juno (118 miles) and Vesta (248 miles). All of them had orbits similar to that of Ceres, and all were extremely small as planets go. More than 1,500 of these *asteroids* have been catalogued, and from photographs taken at Mount Wilson it has been estimated that there are at least 30,000 of them brighter than the nineteenth magnitude. In spite of their number, however, the combined volume of all the known asteroids is probably less than one-twentieth that of the moon.

But Bode's law had a still more important contribution to make. Soon after the discovery of Uranus, it was noticed that

166

the planet failed to follow its predicted orbit exactly. Astronomers decided that another planet beyond Uranus must be influencing its motion. The next term in Bode's series was 388; so a search was on for an unknown planet at a distance of 38.8 astronomical units. Such a planet was found, but not at the correct distance. Neptune's distance is about 30 units instead of the required 38.8. Pluto, the next planet to be found, turned up at 39.5, however, which is in good agreement with the ninth term of Bode's series. So it seems that the "law" is not a law after all. A planet the size of Neptune can hardly be called an "exception."

The failure of Bode's law is a good example of the philosophy that a wrong hypothesis is often better than no hypothesis at all. The "law" started all the excitement about the missing fifth-term planet and helped to stir interest in finding Neptune. These contributions can hardly be called useless. But despite its past usefulness, no known reason for the law has ever been unearthed and it may very well be merely a profound coincidence.

How does a light pipe work?

A rod of glass or of some transparent plastic material can be made to carry light from one end to the other much as a pipe carries water. You may have seen this principle used in window displays and other advertising media. The really unexpected part of all this, however, is the fact that the rod need not be straight. The rod will carry light from one end to the other even if it is bent in the shape of a pretzel—provided the bends aren't too sharp!

This principle produces the spectacular effects seen in brightly illuminated fountains at night. Each column of rising water has within it a beam of colored light—trapped by the smooth surface of the column, but bursting into a myriad of dancing colors as the column breaks into droplets. These and countless other visual experiences can be accounted for by three simple empirical laws which form the basis of the science of *geometrical optics*. But before we get into the subject it should be understood that

167

these laws, like Newton's laws of motion, do not explain *why* light behaves as it does, but merely that it moves about in accordance with certain discernible and predictable rules. These are the law of *rectilinear propagation,* the law of *reflection,* and the law of *refraction.*

The law of rectilinear propagation tells us simply that light travels in a straight line, provided the medium is homogeneous. The laws of reflection and refraction can be explained best in terms of the diagram, Figure 36. Suppose we have a ray of light striking the surface of separation between two transparent substances, such as in the accompanying sketch. One might be air, and the other water. In general, the ray *AO* will split up into a reflected ray *OC* and a refracted ray *OB*. The reflected ray will bounce off and remain within the same substance as the original or incident ray, while the refracted ray will pass through the surface into the second substance. All three rays, however, will end up in the same plane which is perpendicular to the boundary surface between the two media. Another property of the rays, called the law of reflection, states that the *angle of incidence, a,* always equals the *angle of reflection, b.*

$$\text{Angle } a = \text{angle } b$$

These angles are formed between the rays in question and the perpendicular to the surface at point *O*. The law of refraction states that the ratio of the sine of the angle of incidence to the sine of the angle of refraction is a constant.

$$\frac{\sin a}{\sin c} = k \tag{1}$$

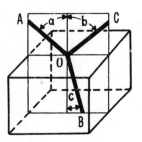

FIGURE 36. A ray of light striking the surface of two transparent media usually divides into a reflected ray and a refracted ray.

This equation tells us that the ratio above remains constant for the two media and does not depend upon the angle of incidence.

Inspection of equation (1) can tell us that a refracted ray will be brought closer to the perpendicular or farther from it depending upon whether k is larger or smaller than 1. If k is larger than 1, then sin a must be larger than sin c. If this is true, a must be larger than c. So if k is larger than 1, the refracted ray must be closer to the perpendicular than the incident ray. If k happens to be smaller than 1, the unexpected sometimes happens. Sin a must then be smaller than sin c, and, therefore, a must be smaller than c. If this is true, then the refracted ray must be farther from the perpendicular than the incident ray. When a has such a value that sin $a = k$, then sin c must equal 1. But if sin $c = 1$, then $c = 90°$. The refracted ray is then sent grazing off parallel to the boundary surface. For a slightly larger angle of incidence, there is no refracted ray at all, and all the light is reflected—as with a mirror. This situation is called *total reflection.*

Beams of light entering a plastic rod or a column of water at a sufficiently small angle are totally reflected from the internal walls of the material. They remain trapped within the light conducting substance, bouncing from wall to wall, until they reach a point where equation (1) can be satisfied. Part of the light is then able to escape and illuminate droplets of water or any other objects in its path.

What are exponents?

Exponents—the small numbers written at the upper right-hand portion of a base number—are merely mathematical short cuts used to compress long numbers. The number 5^2, for example, merely means 5×5, or two 5's multiplied together. Similarly, $6^4 = 6 \times 6 \times 6 \times 6$; $10^6 = 10 \times 10 \times 10 \times 10 \times 10 \times 10$; and in general, n^a equals the result of n multiplied by itself a times. In speaking of such numbers, we say that n is raised to the a power. When the exponent is 2 or 3, however, we fall back on ancient Greek usage and speak of n squared or n cubed.

Thus far we have considered the exponent to be an integer,

169

but what happens when the exponent is zero? Take 4^0 for example. To find the meaning of this expression we must determine the value of the exponential 4^a as a approaches zero. We know from our study of arithmetic that $\sqrt[2]{4}$ is often written $4^{\frac{1}{2}}$. That is, the degree of the root to be taken is written in the form of a fractional exponent. Then

$$
\begin{aligned}
\sqrt[2]{4} &= 4^{\frac{1}{2}} &= 2 \\
\sqrt[4]{4} &= 4^{\frac{1}{4}} &= 1.414 \\
\sqrt[8]{4} &= 4^{\frac{1}{8}} &= 1.189 \\
\sqrt[16]{4} &= 4^{\frac{1}{16}} &= 1.089 \\
\sqrt[32]{4} &= 4^{\frac{1}{32}} &= 1.044 \\
\sqrt[64]{4} &= 4^{\frac{1}{64}} &= 1.022 \\
\sqrt[128]{4} &= 4^{\frac{1}{128}} &= 1.011 \\
\sqrt[256]{4} &= 4^{\frac{1}{256}} &= 1.005
\end{aligned}
$$

If we were to carry this process on indefinitely, we would find that the answer gets closer and closer to 1 as the fractional exponent gets smaller. In mathematical terms, the exponential *approaches* 1 as a limit as the exponent *approaches* 0. From this we deduce that $4^0 = 1$. While we have illustrated this fact for the number 4, the same reasoning will show that any number raised to the zero power is $1: n^0 = 1$.

It's also possible to have exponentials where the exponent is a mixed number such as $4^{1\frac{1}{2}}$. This converts readily to $4^{\frac{3}{2}}$ which tells us to perform two operations. First we must take the square root of 4; then we must cube that number. Or we can do it the other way around without affecting the final result. This comes from the fact that

$$ 4^{\frac{3}{2}} = (4^3)^{\frac{1}{2}} = \sqrt[2]{4^3} = \sqrt{4^3} $$

(When no numeral appears above the radical sign, a 2 is understood; that is, $\sqrt[2]{4}$ is written $\sqrt{4}$.)

In addition to having zero, fractions, and mixed numbers as exponents, it's also possible to have negative exponents. Numbers with negative exponents, as n^{-a} are the reciprocals of the same numbers with positive exponents: $\frac{1}{n^a}$.

170

How did the musical scale evolve?

Strike any two notes on a piano and even an accomplished musician will have some difficulty in determining the interval or number of notes between the two. But the octave is quite another matter. Strike any two notes an octave apart and the feeling of similarity is striking to anyone that is not completely tone-deaf. Many classical and medieval writers wondered about the mathematical relationships between notes, but it was not until the middle of the nineteenth century that the mystery of the octave was solved.

In 1863, Helmholtz pointed out that most instruments do not emit a "pure" tone when a note is played. On the contrary, a musical note usually consists of a complex sound made up of a *fundamental tone* accompanied by *harmonic overtones* whose frequencies are 2, 3, 4, 5, . . . times that of the fundamental. Suppose a certain note has a fundamental frequency of 100 vibrations per second. Its overtones will be $2 \times 100 = 200, 3 \times 100 = 300, 4 \times 100 = 400$, and so on. The compound sound consists of

100   vibrations per second (fundamental tone)

$\left. \begin{array}{l} 200 \\ 300 \\ 400 \\ 500 \\ 600 \\ \text{etc.} \end{array} \right\}$ vibrations per second (overtones)

If a note 1 octave higher in pitch is played (200 vibrations per second), the complex sound consists of

200   vibrations per second (fundamental tone)

$\left. \begin{array}{l} 400 \\ 600 \\ 800 \\ \text{etc.} \end{array} \right\}$ vibrations per second (overtones)

A comparison of the two sets of frequencies shows that the second group of overtones coincides with the alternate overtones of the first note.

171

First note	Second note
100	
200	200
300	
400	400
500	
600	600
700	
800	800
etc.	etc.

When the second note is played, the ear once more hears a part of the original sound and we have a subjective feeling that the notes are somehow related by the elements common to both.

This explanation tends to be borne out by the comparative difficulty experienced in judging the interval of an octave between notes played consecutively on the flute, which produces feeble overtones, or at the high treble end of a piano where the overtones are reaching the upper limit of sensitivity of the ear.

A similar, though less marked, similarity can be noticed for the *fifth*, the interval between two notes whose frequencies are in the ratio $\frac{3}{2}$. The fifth is so named because the higher note is just five notes above the lower on the scale we use. As with the octave, the kinship associated with the notes of a fifth results from overtones that are common to both notes.

If we let the frequency of C equal 24, then G, its fifth, will have a frequency of $\frac{3}{2} \times 24 = 36$. The overtones of each note are as follows:

C	G
24	36
48	72
72	108
96	144
120	180
144	216
168	252
192	
216	
etc.	etc.

It can be seen that alternate overtones of G correspond with every third overtone of C. If we denote the octave of C as C_1, it's possible to begin the construction of musical scale based on the octave and the fifth.

$$
\begin{array}{ccc}
C & G & C_1 \\
24 & 36 & 48
\end{array}
$$

In this scale, G is a fifth above C and its frequency is three-halves of C; C_1 is an octave above C and its frequency is twice that of C.

The next step in the construction of a musical scale is the selection of a note F such that C_1 is a fifth above it. In other words,

$$F \times \tfrac{3}{2} = C_1$$

$$F \times \tfrac{3}{2} = 48$$

$$F = \tfrac{2}{3} \times 48 = 32$$

The revised scale is then

$$
\begin{array}{cccc}
C & F & G & C_1 \\
24 & 32 & 36 & 48
\end{array}
$$

But this new scale now contains a new interval, the fourth, consisting of C–F or G–C_1. The ratio corresponding to the fourth is

$$\frac{F}{C} = \frac{32}{24} = \frac{4}{3}$$

or,

$$\frac{C_1}{G} = \frac{48}{36} = \frac{4}{3}$$

The notes of a fourth are not too readily recognizable as a "natural" interval because the common overtones are quite high in frequency and are not usually too prominent in musical instruments.

The note D can be added by taking a fourth below G. Since $D \times \tfrac{4}{3} = G = 36$, the new note is $D = 27$; and the scale becomes

$$
\begin{array}{ccccc}
C & D & F & G & C_1 \\
24 & 27 & 32 & 36 & 48
\end{array}
$$

Similarly, if A is taken a fifth above D, E a fourth below A, and B a fifth above E, the final scale is

C	D	E	F	G	A	B	C_1
24	27	$30\frac{3}{8}$	32	36	$40\frac{1}{2}$	$45\frac{9}{16}$	48

This scale contains only fourths and fifths in addition to the octave and is said to be unrivaled for melodic purposes. It was discovered empirically by the ancient Greek mathematician, Pythagoras, and endured until the beginning of the eighteenth century when the new art of harmony caused the alteration of E, A, and B to slightly lower frequencies.

The idea of harmony—the playing of two or more notes simultaneously to form a chord—is so familiar that we might expect the practice to be extremely old. And, in fact, a few ancient Egyptian drawings do show instruments being fingered in a way to suggest harmony. It seems probable, however, that the practice was not widespread since the principles of harmony were most certainly unknown to the ancients. There is no unequivocal evidence either way, however, and the question is indeed a perplexing one.

In modern harmony, the rules for obtaining agreeable combinations are well known. When two notes are sounded together, there may be a peculiar and disagreeable effect consisting of a throbbing or beating of the notes. The effect can best be understood with the help of a hypothetical experiment. Imagine two sources of musical notes; one fixed in frequency, and the other capable of being changed continuously. When the two tones are in unison, they blend together and we hear a single tone. If the second tone is gradually increased in pitch, *beats* are heard which become more and more rapid and finally die away. With the beats comes an unpleasant *discord* which is strongest at about 25 beats per second, then diminishing and finally disappearing at higher frequencies.

Beats may be produced by fundamental tones or by any combination of their overtones that happen to be close enough in frequency. When a Pythagorean third $(C-E)$ is played, a marked dissonance is heard resulting from the interaction of the overtones.

174

By changing the frequency of E to 30 vibrations per second, the modern third is made more harmonious. Because of this change, the fourth and fifth above E are changed to 40 and 45 respectively to produce the true diatonic scale.

C	D	E	F	G	A	B	C_1
24	27	30	32	36	40	45	48

An examination of this scale shows the intervals between adjacent notes to consist of the following ratios:

Interval	Interval ratio
Major tone	$\frac{9}{8}$
Minor tone	$\frac{10}{9}$
Semitone	$\frac{16}{15}$

The major difficulty with this scale lies in the variable interval ratios. If we were to construct a second scale using the octave $D–D_1$, the frequency of E would be $\frac{9}{8} \times 27 = 30\frac{3}{8}$—a frequency somewhat higher than that of E on the original scale. In order to accommodate any number of scales, a musical instrument would require a multiplicity of different frequencies for each note. The so-called *even-tempered* scale is used just to circumvent this difficulty.

If the scale is expanded to include the 5 sharps and flats, there are 13 notes or 12 intervals to an octave. If we denote the ratio for the interval as i, then $C\sharp = C \times i$. Similarly, $D = C\sharp \times i = C \times i^2$; $D\sharp = D \times i = C \times i^3$, and so on. Each note is equal to i times its predecessor. The octave, then, is $C_1 = C \times i^{12}$ or

$$i^{12} = \frac{C_1}{C} = 2$$

Solving this equation gives $i = 1.06$—the interval between half tones on the even-tempered scale.

How much do metals expand when heated?

Anyone that has walked along railroad tracks has noticed the cracks between successive rails. It would have been much simpler, of course, to place one rail directly against the next leaving no gap at all between them. This can't be done because expansion of the steel on a hot day might cause the rails to buckle. It's important, then, to know just how great this gap should be. It should be large enough to prevent buckling on the hottest day, yet small enough to prevent excessive wear to train wheels. The optimum gap size can be determined only through a quantitative knowledge of the expansion of steel.

In working with solids we are usually concerned with linear expansion. Of course, metals expand in all directions, but the nature of the problem usually confines our attention to one most significant direction. The change in height or width of train rails is really not important—it's the change in length that we are interested in.

The expansion of a material is measured by a quantity called the *coefficient of linear expansion*. This is the fraction of itself that a length of a substance expands when the temperature is raised 1° (usually centigrade). Suppose that a rod of metal is 1 unit in length. In addition, suppose that an increase in temperature of 1°C results in an increase in length of one-millionth of a unit. The coefficient of linear expansion is then 1 part per million, per degree centigrade. It is usually written 1×10^{-6} per degree centigrade. The coefficients of linear expansion of a few substances are given below.

Substance	Coefficient of linear expansion
Aluminum	22×10^{-6}
Brass	18×10^{-6}
Brick	9.5×10^{-6}
Copper	16×10^{-6}
Iron	11×10^{-6}
Mercury	30×10^{-6}
Invar (a nickel-iron alloy)	0.9×10^{-6}

If an iron rail is L units in length, the change in length per degree centigrade is equal to $L \times a$, where a is the coefficient of linear expansion. If such a rail is subjected to a temperature difference of $T_1 - T_2$, the total expansion becomes $L \times a \times (T_1 - T_2)$. Let's assume a maximum temperature change between summer and winter of 100° C. The change in length of a mile of iron rail (let's say 5,000 feet) will then be

$$(5,000)(11 \times 10^{-6})(100) = 5.5 \text{ feet}$$

The change in length to be expected in such a length of track is 5.5 feet.

What are "radio stars"?

Most of what man has been able to deduce about outer space has been garnered through a rather narrow range of wavelengths that can penetrate the earth's atmosphere. The most important of these wavelengths are known as the visible spectrum. Under good atmospheric conditions, and at night, the human eye can observe the radiations from the stars that happen to fall in this visible spectrum. This range has been extended somewhat through the use of thermopiles and bolometers which extend our knowledge into the infrared region at one end, and other instruments such as photoelectric cells that enable us to "see" into the ultraviolet region. But even with these aids, the atmosphere is "transparent" only for a relatively narrow range of wavelengths. Fortunately, however, there is another atmospheric "window" in the region of radio waves. This window extends from wavelengths of 0.25 centimeter on the short-wave side to perhaps 20 meters on the long-wave side. These are the wavelengths used by man for the many communications and radar applications that we know so well today. Only recently have these wavelengths been considered important for astronomical observation.

In December, 1931, Karl G. Jansky made the surprising discovery that radio waves were reaching the earth from outer space. Over the next few years he concluded that most of the energy was coming from the direction of the center of the galaxy with the source distributed along the Milky Way. Although Jansky's discovery was to give birth to a new science, radio astronomy, it was not fully appreciated at the time. We know today that the use of radio waves has given to astronomy a completely new avenue for the exploration of the Galaxy and all space.*

Some sixteen years later, Hey, Phillips, and Parsons discovered that the noise coming from the region of *Cygnus* seems to originate in a high intensity source having an angle of perhaps 8′ of arc. But even more amazing, this source was in a region of the galaxy completely devoid of any outstanding visible objects. Shortly afterward, Bolton gave an account of several other point sources of radio energy. Many such "radio stars" are known today, but the majority show no connection with nearby visible objects. This suggests the possibility that they may be undiscovered types of stars of very low brilliance but producing intense radiation at radio wavelengths. One theory holds that most of the galactic radiation at radio wavelengths is a result of point sources such as radio stars. According to the theory, the density of such radio stars would be approximately 3 per cubic parsec** which corresponds to a separation of the sources of 2×10^{18} centimeters—a figure which is comparable to the separation of the visible stars. It is also pointed out that the ratio of the intensity of the most powerful radio star to that of the whole sky is of the same order of magnitude as the ratio of the intensity of the brightest visible star to the total emission of all the other stars.

If this reasoning is correct, and it is by no means universally accepted, there are about as many radio stars in the Galaxy as visible ones—perhaps 100,000,000,000! If so, these stars can be "seen" only through the use of the radio telescope.

Why do certain meteor showers recur annually?

In addition to the great number of sporadic shooting stars which are seen throughout the year, there are occasional occurrences of meteor showers. At such times, a certain section of the sky provides many more than the usual number of bright streaks. Many such showers recur with remarkable regularity on the same

* The word *Galaxy* has come to be used in reference to the large group of stars of which the sun is a member, as well as to other similar but distant systems—the external galaxies.
** One cubic parsec $= 13.6 \times 10^{40}$ cubic miles.

dates from year to year. Measurements made on such meteors indicate that all are traveling in parallel paths, and with identical speeds in their journey through space. Astronomers conclude, therefore, that the shower is a result of a swarm of small pebbles moving in almost identical orbits around the sun. Moreover, the meteoric material seems to be fairly evenly distributed around the orbital path, forming a dense ring of pebbles that follow each other around the sun. Measurements made on the direction and velocity of such meteors show that they follow an elongated orbit similar to those of elliptic comets.

The orbit of the earth is so situated that in a number of cases it intersects the orbit of a meteor stream. Each time the earth reaches this point in its orbit a meteor shower takes place. The showers may last for periods varying from a few hours to several days, during which time the meteor rate increases in varying amounts up to 50 or 100 per hour. Some of the more important annual showers are given below.

Shower	Date of maximum	Normal hourly rate
Quadrantids	January 3	35
Lyrids	April 21	8
Aquarids	May 6	12
	July 28	10
Perseids	August 10–14	50
Orionids	October 20–23	15
Taurids	November 3–10	10
Leonids	November 16–17	12
Geminids	December 13–14	60
Ursids	December 22	13

One of the greatest meteor swarms is known as the *Giacobinids* because it is believed to be associated with the Giacobini-Zinner comet of 1900. Whenever the earth crosses the comet's orbit *close to the comet,* a spectacular shower takes place reaching 300 or 400 meteors per hour. At other times the crossing takes place either long before or long after the comet and few meteors are seen. Evidently the meteoric debris of the comet is rather closely grouped. According to one theory, this material will eventually spread out over the entire orbit as a result of perturbation produced by the gravitational forces of the planets. The Giacobinids (October 9) will then produce yearly showers as do the more reliable meteor streams listed earlier.

Where do shooting stars come from?

One of the great controversies among astronomers has been concerned with the origin of shooting stars or meteors. Generally speaking, it's agreed that those meteors producing recurrent showers must move in closed orbits around the sun. These are accepted as members of the solar family. But how about the sporadic meteors—those not contained in any recognized stream? These constitute a large portion of all visible meteors; and many astronomers held that they must be of interstellar origin—visitors from outer space.

The use of the techniques of radio astronomy seems to have solved this problem once and for all; practically all meteors belong to the solar system, and the small number in doubt probably do also.

All members of the solar system move in closed elliptical paths around the sun. The velocity of such a solar member is given by

$$V^2 = Mg \left(\frac{2}{R} - \frac{1}{a} \right)$$

where M is the mass of the sun, g is the gravitational constant, R is the radius vector, or distance from the body to the sun, and a is the semimajor axis of the earth's orbit. (This relation assumes that the mass of any such body is small compared to the sun's—a reasonable assumption.)

If the meteor is from outer space, however, it will move in a hyperbolic orbit and its velocity will be *greater*.

$$V^2 = Mg \left(\frac{2}{R} + \frac{1}{a} \right)$$

Notice that the two equations above differ only in the element $\frac{1}{a}$ within the parentheses. The intermediate condition when

180

$$V^2 = Mg\frac{2}{R}$$

is of particular importance because it corresponds to the critical case of a meteor moving in a parabolic orbit. The accompanying diagram illustrates the significance of these particular curves (Figure 37). Although they have been drawn to appear quite similar, a meteor traveling in the outer (hyperbolic) orbit would leave the

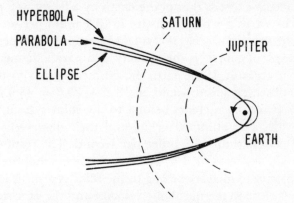

FIGURE 37. Three possible orbits of meteors drawn with respect to the orbits of the earth, Jupiter, and Saturn.

solar system altogether, never to return, while a body moving in the innermost (elliptical) orbit would return after some hundreds of years. The intermediate parabola is the critical curve between the ellipse and the hyperbola. Using the equations given above, the critical speed corresponding to a parabolic orbit can easily be calculated. Then all that need be done is to measure the velocities of many meteors and compare these figures with the critical parabolic velocity. Those meteors exceeding the critical speed must have hyperbolic orbits and, consequently, must come from outer space. Those having lower speeds must move in elliptical orbits as *bona fide* members of the solar system.

If we assume that the earth's orbit is a circle having a radius equal to *a*, the velocity of the earth becomes

181

$$V_e^2 = \frac{Mg}{a}$$

Similarly, the critical parabolic velocity is

$$V_p^2 = \frac{2Mg}{a}$$

From these two equations, it's evident that the velocity of a parabolic meteor exceeds that of the earth by a factor $\sqrt{2} = 1.414$. Since the earth's orbital velocity is 18.2 miles per second, the parabolic velocity $V_p = (18.2)(1.414) = 25.6$ miles per second. This speed, of course, is the velocity of a parabolic meteor with respect to the sun. If the earth should meet such a meteor head on, the observed speed would be $18.2 + 25.6 = 43.8$ miles per second. If all the meteors belong to the solar system, then all should travel in elliptical orbits and their observed velocities should be less than 43.8 miles per second. The results of such measurements performed by radio astronomers indicate that the vast majority of meteors belong to the solar system. A few travel at speeds close to the parabolic velocity and the precision of the measurements is not sufficient to make a definite determination. It is estimated, however, that interstellar meteors, if they exist, constitute less than 1 per cent of all visible meteors.

Men have wondered about the origin of meteors for thousands of years. They need wonder no longer; a new branch of science has now given the answer.

How fast do meteors travel?

Although a visual observer may see only a few meteors, or shooting stars, per hour, it is estimated that at least 8 billion enter the atmosphere every 24 hours! Most of these are merely faint telescopic objects, however, and quite invisible to the unaided eye. The advent of any particular meteor is completely unpredictable and the entire display is over within a fraction of a second. It's no wonder, then, that the accurate determination of their speeds has presented an enormously difficult problem to astronomers. Until recently, the only completely acceptable method of measuring meteoric speeds has been to photograph the same meteor from two locations on the earth's surface. Each photograph accurately positions the meteor trail against the background of stars. In order to determine the duration of the meteor track, the shutter of the camera is closed for a short period of time at regular intervals of about one-twentieth of a second. The meteor trail is thereby broken up into segments resembling a dashed line. Counting the segments gives the time duration of the trail, and trigonometry gives its length; so the velocity is computed by dividing length by time. Careful measurements of this kind on a very few bright meteors indicate an observed average speed of about 22 miles per second with respect to the earth.

While this method, perfected by F. L. Whipple of Harvard, gives precise measurements, it is limited to very bright meteors. Unfortunately, bright meteors are relatively scarce and, of course, none can be detected during the daylight hours, or when the sky is obscured by clouds. A new method had to be found, and the new science of radio astronomy provides several methods which have revolutionized the measurement of meteor speeds. As a result of these new techniques, we know that the great majority of meteors have speeds between 9 and 42 miles per second.

When a meteor burns away in the high atmosphere, it leaves behind a brilliant streak of light which we normally see, and, in addition, a dense column of *ionized* air and meteor material. The great heat produced by friction temporarily dislodges electrons from affected molecules resulting in the production of electrons and *ions*. Matter in this state is said to be ionized. Instead of electrically neutral molecules, the column of ionized air contains electrically charged particles which are capable of reflecting radio waves. This makes it possible to obtain radio "echoes" from meteor trails by employing well-known radar techniques.

The principle is easy to follow. Imagine a burst of radio waves —a *pulse* in the jargon of radar—that is directed toward a sector of the sky to be studied. The speed of this pulse is equal to that of light, 186,000 miles per second. If the distance to the meteor trail is R miles, the time required for the round trip is $2R \div 186,000$ seconds. Electronic equipment, similar to the radar sets of World War II, measures the round-trip time to the ionized meteor trail with considerable accuracy and converts the time interval to distance. Most meteor trails occur at heights of 40 to 80 miles, so the average time interval to be measured is 0.0006 second or about 0.6 millisecond. (1 millisecond $= 10^{-3}$ second.) This technique, known as radar ranging, provides the distance to the meteor trail. To find the velocity of the meteor, one additional effect must be taken into account. This effect causes the *strength* of the echo to vary from one instant to the next in a definite and predictable manner. An explanation of the Fresnel diffraction principle, as it is known, is beyond the scope of this book, but it can be shown that the echo signal will alternately increase and decrease in magnitude until the meteor is completely disintegrated and the ionized column disappears. The velocity of the meteor is found to be

$$V = \frac{0.3\sqrt{\lambda R}}{T}$$

where λ is the wavelength of the radar signal, R is the distance to the meteor, and T is the time interval between the first maximum and the first minimum of the received echo signal. To get an idea of the magnitudes involved, let's assume we are dealing with a meteor traveling at 25 miles per second at a height of 60 miles. If the wavelength in use is 30 feet long, the time interval T becomes

$$T = \frac{0.3\sqrt{\dfrac{(30)(60)}{5,280}}}{25}$$

$$T = 0.007 \text{ second} = 7 \text{ milliseconds}$$

The degree of precision needed in such meteor "speedometers" is indeed great.

Radio astronomy, through methods such as these, has been able to provide with ease many accurate measurements that previously had been practically impossible. Of course, daytime detection of meteors can be accomplished *only* by radio techniques, and daytime meteor streams have actually been discovered in this way.

An unexpected by-product of this research has made possible one of the most improbable means of communication ever devised by man—meteor burst communication. In such a system, radio waves are transmitted over long distances by bouncing them off meteor trails! The techniques are highly developed even today. In a practical system, the intelligence to be transmitted is not ordinary speech, but high-speed telegraphic signals. The transmitter waits until a favorable meteor trail begins and then sends out a *burst* of signals, from which comes the name of the technique. Since there are so many meteors in the sky at all times, the transmitter does not have to wait too long for a favorable one to occur.

What's the probability of drawing a flush?
A flush is a poker hand consisting of 5 cards all the same suit. To calculate the probability of being dealt a flush, it's necessary to determine the probability of getting each of the required cards and then compute the product of their separate probabilities.

When one is dealt a flush, it doesn't matter in the least which card he begins with. There are 52 cards in a poker deck and any one of the 52 will do. So the probability of getting the "right" first card is $\frac{52}{52} = 1$.

At this point, there are 51 cards unaccounted for, but *only 12* will be acceptable, since the second card must be of the same suit as the first. The probability of getting 1 of the 12, then, is $\frac{12}{51}$. In the same way, the probabilities corresponding to the remaining cards are $\frac{11}{50}$, $\frac{10}{49}$, and $\frac{9}{48}$. So the probability of getting all five cards the same suit is

$$1 \times \frac{12}{51} \times \frac{11}{50} \times \frac{10}{49} \times \frac{9}{48} = 0.0022$$

or about 1 chance in 500. The probabilities of being dealt other poker hands are determined in another section of the book.

185

If you're playing "draw" poker, you can use the rules of probability to determine your chances of "filling a straight," or drawing a third card to a pair. Suppose you have a hand consisting of *x,* 6, 7, 8, and 9, and you want to calculate the probability of completing the straight. In this example, either a 5 or a 10 will do. There are $52 - 5$, or 47, cards whose locations are unknown to you. Of these, the card that you want must be either a 5 or a 10. Since there are 8 favorable possibilities out of 47, the probability of drawing *either* a 5 or a 10 is $\frac{8}{47} = 0.17$, or 17 chances in 100. Notice that trying to fill a straight such as 6, 7, 9, 10 is twice as difficult (probability $= \frac{4}{47}$) since only 8's can be successful. "Never draw to an inside straight" has become axiomatic.

Many modern card games purporting to be variations of poker use a multiplicity of jokers and wild cards (often including, unbelievably, the split-whiskered kings and one-eyed jacks). The serious student of poker should avoid these monstrous games at all cost, since he will undoubtedly lose all his funds well before the rules of probability have had a sufficiently large statistical sample on which to operate. All that the theory of probability can tell you is that you will *probably* be dealt 1 flush in 500 hands. It makes no promises. You may get a flush on the first deal—and you may not get 1 in 10,000 deals. But given a large enough sample, a flush will turn up in just about the "right" number of hands.

Calculations like those given above show that the accepted seniority of the various poker hands corresponds precisely to their order of mathematical probabilities. It seems unlikely that any mathematicians were consulted; so we can only assume that the game evolved "correctly" on an empirical basis as countless millions of players risked their money in millions of poker games all over the world.

How are time-exposed photographs of stars made?

In ancient times, man discovered that there is a regularity to the apparent motion of all fixed stars. Specifically, they seem to rotate at the same rate in circular arcs at *right angles* to the earth's polar axis. For practical purposes, the plane in which a star appears to move is perpendicular to the line joining your eye to the North Star. The angle of this line above the horizon also happens to be the latitude of your location; so this angle will become greater as you move toward the north and it will be smaller as you move south. But for any one location, the line between the observer and the North Star always makes the same angle with the horizon. Similarly, the angle between this line and any star is constant throughout the night. This makes it possible to take time exposures of the stars as they rotate across the nighttime sky. Here's how it's done (Figure 38). A shaft is pointed directly at Polaris, the North Star, and fixed in that position. A pipe is then slipped over the shaft and left free to rotate. A telescope is mounted on the pipe. The angle between the telescope and the pipe is adjusted so that the desired star is in view. The

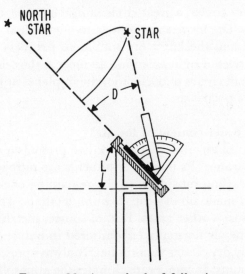

FIGURE 38. A method of following a star in its course through the sky. Angle *L* is the observer's latitude and angle *D* is the star's declination.

187

telescope is then clamped to the pipe to preserve that angle. We can now follow the course of our star throughout the night merely by rotating the telescope on its free axis about the shaft without raising or lowering it. If we now add a clockwork mechanism that rotates the telescope through 360° in a day, it will point toward the same star long enough for us to take a time photograph. The time used in this arrangement is a sidereal day, or the time elapsed between two successive appearances of the same star directly overhead.

This regularity of movement on the part of the fixed stars suggested to the early astronomers a method of mapping their location in the sky. If any two stars are selected, the observer will note that one will cross the meridian (i.e., pass directly overhead) exactly the same length of time before or after the other star each night. This means that the stars are fixed with respect to each other, and so can be given positions in the great celestial sphere of the heavens. This is done by noting the crossing of two imaginary circles, a great circle of *right ascension,* and a circle of *declination.* The former are similar to our ordinary meridians of longitude and the latter are similar to parallels of latitude. The great circles of right ascension all meet at the celestial poles, while the small circles of declination lie in planes at right angles to the earth's polar axis.

What is the water content of the air?

The ocean of air under which we live presents a deceptively simple appearance. Its major constituents are nitrogen, oxygen, argon, and carbon dioxide with the remaining one-hundredth of 1 per cent made up of neon, helium, methane, krypton, and traces of a dozen other gases. But, of course, this description is of *dry* air. The air normally encountered in nature is rarely, if ever, perfectly dry; water vapor is nearly always present in small, varying amounts. In temperate climates the atmosphere's water content varies from just about zero when dry winds are blowing, to perhaps 3 per cent by volume on humid summer days. In tropical climates the water content may be as high as 4 per cent, which is about the upper limit. To make room for this small but important component of the air, the quantities of the other gases must change correspondingly. The table below gives the composition of moist air for varying amounts of water-vapor content.

Water vapor	Nitrogen	Oxygen	Argon	Others
0%	78.08%	20.95%	0.93%	0.04%
1	77.30	20.74	0.92	0.04
2	76.52	20.53	0.91	0.04
3	75.54	20.32	0.90	0.04
4	74.96	20.11	0.89	0.04

So it appears that the air's motion is not its only unpredictable characteristic. Its very make-up changes considerably from one day to the next.

The air may contain, on the average, about 2 per cent of water vapor by volume; yet this minute amount of water is responsible for the whole gamut of climatological variation from arid desert to tropical paradise. Every drop of fresh water has found its way from the ocean's surface to the atmosphere in order to fall as rain or snow on some part of the earth. This amazing and extremely important function of the atmosphere is little appreciated by most of us who much prefer fair weather to foul.

While the moisture content of the air can be described as above in terms of per cent content by volume, such a method doesn't reflect adequately the way in which moisture affects our comfort and well being. A more appropriate concept from the human point of view is that of *relative humidity*—the degree to which air is saturated with moisture. Perfectly dry air has a relative humidity of 0 per cent, while "soggy" air that will accept no additional moisture has a relative humidity of 100 per cent. The intermediate amounts of relative humidity can be explained with the help of a new concept—*vapor pressure.*

The atmosphere is made up of many gases, each of which makes its contribution to the total atmospheric pressure. The

189

way in which the individual pressures add up is exceedingly simple, yet quite unexpected, even when we know the trick. The principle involved is known as Dalton's law of partial pressures. The law states that the pressure exerted by a mixture of gases (like air) is equal to the sum of the individual pressures that each gas would exert if it alone occupied the entire volume. To illustrate, suppose we have a quantity of gas #1 in an otherwise empty bottle and suppose the pressure within the bottle is p_1. Similarly, imagine that a second bottle, identical to the first, contains an amount of gas #2 which produces a pressure of p_2. Dalton's law states that if both quantities of gas were in one bottle the resulting pressure, p, would be

$$p = p_1 + p_2$$

Since there may be many such pressures, the general formula becomes

$$p = p_1 + p_2 + p_3 + \cdots$$

where p_1, p_2, p_3, . . . represent the individual or partial pressures exerted by each gas in the mixture. Since the atmosphere consists of many gases, the total atmospheric pressure is the sum of the partial pressures of the various gases. Water vapor, which is usually present in air, is a true gas and adds its contribution to the total. The partial pressure of water vapor varies from almost 0 to as much as 4 per cent of the total atmospheric pressure.

The maximum amount of water vapor that the atmosphere is capable of containing depends only upon the temperature, and not upon the other gases present. Under these conditions, the atmosphere is said to be saturated with water vapor and the corresponding partial pressure is called the saturation vapor pressure at that particular temperature. The table below gives the saturation vapor pressures for several corresponding temperatures.

Temperature, °F	Saturation vapor pressure, pounds per square inch	Per cent moisture by weight
0	0.076	0.083
10	0.13	0.14
20	0.21	0.22
30	0.34	0.34
40	0.51	0.51
50	0.74	0.72
60	1.1	1.0
70	1.5	1.4
80	2.1	2.0
90	2.9	2.6
100	3.9	3.5

The physical concepts discussed above have been included to emphasize the fact that the vapor pressure of water exists as a separate entity—entirely independent of the other gases in the air. Its pressure, when added to the partial pressures of all the other atmospheric gases, constitutes the total atmospheric pressure.

We are now in a position to define relative humidity. Relative humidity is simply the ratio of the actual vapor pressure to the saturation vapor pressure at a particular air temperature.

$$\text{Relative humidity} = \frac{\text{vapor pressure}}{\text{saturation vapor pressure}}$$

Relative humidity is particularly appropriate as a measure of the air's moisture content because it is related to the effect that moist air has on people. If the relative humidity is close to 100 per cent, perspiration does not evaporate readily and we feel hot and uncomfortable even if the temperature is not excessive. If the relative humidity is low, on the other hand, rapid evaporation takes place and we feel cool even at temperatures well above the "normal" 72°F.

Those of us living in the temperate zone in heated homes know that the air is drier in winter than in summer. This can be explained easily with the help of the table of partial pressures given above. Suppose the outside air is saturated (100 per cent relative himidity) at a temperature of 20°F. What will be the relative humidity inside the home if the air temperature is 80°F? The vapor pressure of the outside air is 0.21 pound per square inch. This is enough to account for a relative humidity of 100 per cent. But the heated air inside the home requires 2.1 pounds per square inch or 10 times that vapor pressure to reach saturation. The relative humidity of the inside air is then about one-tenth as great, or about 10 per cent. No wonder it feels dry!

If you are curious about the relative humidity in your home, you can measure it in a few minutes with a simple homemade *psychrometer* consisting of two thermometers, a glass of water, and a piece of gauze. Here's how its done. The two thermometers are arranged side by side so that one measures the normal air temperature and the other, the *wet-bulb temperature*. The latter quantity is the air temperature indicated by a thermometer whose bulb is covered by a thin, wet cloth *during brisk ventilation*. The cloth is wrapped around the bulb and the other end placed in the water to provide a short capillary path for the water from glass to thermometer. The wet-bulb temperature is related to the humidity because the drier the air the greater the cooling by evaporation, and the lower the wet-bulb temperature. The number of degrees difference between the dry-bulb and wet-bulb readings is called the wet-bulb depression. At 65°F, for example, a wet-bulb depression of 13° corresponds to 40 per cent relative humidity. The relationship between temperature, wet-bulb depression, and relative humidity is a very complicated one that has been determined by experimentation and made available in the form of *psychrometric tables* and charts. One such table is published by the United States Weather Bureau, and textbooks on thermodynamics usually have the information in chart form.

The simplified table given below will enable you to convert the readings of your homemade psychrometer to relative-humidity figures with reasonable accuracy. The data are presented for several room temperatures most likely to be found in the home.

Relative humidity	Wet-bulb depression for a dry-bulb temperature of				
	65°F	72°F	80°F	90°F	100°F
0%	24.0	27.5	31.8	37.3	43.3
10	21.0	24.0	27.4	32.0	36.6
20	18.3	20.7	23.5	27.1	30.7
30	15.6	17.5	19.8	22.7	25.4
40	13.0	14.6	16.4	18.8	20.8
50	10.6	11.9	13.2	15.1	16.6
60	8.3	9.3	10.4	11.6	12.9
70	6.0	6.8	7.6	8.4	9.4
80	4.0	4.4	4.8	5.5	6.0
90	2.0	2.2	2.3	2.9	
100	0	0	0	0	

To use the tables merely select the column corresponding to the temperature of the dry-bulb thermometer—or the closest column, if necessary. Move down the column until you find the number equal to or close to the wet-bulb depression. Then move to the left and read the relative humidity in the extreme left-hand column. You can get more exact answers, of course, by referring to a psychrometric table which contains this information in greater detail.

Do the stars move?

Aristotle taught that the heavens were immutable and for 2,000 years men believed him. Then in 1718 Edmund Halley, the astronomer who predicted the return of the comet that bears his name, explained that the stars did really move. He found that Sirius and a few other bright stars had moved as much as the apparent diameter of the full moon from the locations assigned to them by Ptolemy. Since then it has been determined that practically all the stars are moving, although the more distant nebulae are moving so slowly that their positions can serve as a backdrop to show the motion of the nearer ones.

The *proper motion* of a star is the angular change of its position in the heavens. The swiftest is known as Barnard's star after the astronomer who first noted its rapid flight. It moves across the

193

heavens at the rate of 10.3 seconds per year, or an amount equal to the moon's apparent diameter in 175 years. The average motion for those stars visible to the naked eye is about 0.1 second per year.

We usually think of the North Pole Star as if it were unchangeable and always located at the celestial pole. As a matter of fact, it's a fortunate historical accident that there happens to be a bright star near the North Pole in our own time. When the Pyramid of Cheops was built, another star, alpha in the constellation Draco, was located near the North Pole, about three degrees below the true celestial pole. Two shafts in the Pyramid were so located that light from this star shone directly into the inner chambers. Between 2000 B.C. and about 1000 A.D. there was no bright star near the Pole. Today our own Pole Star, Polaris, is within 1°, or about two moon breadths, of the Pole itself. It can be found at the end of the Little Dipper's handle. There is no such bright star near the South Pole.

What is meant by reductio ad absurdum?

Many classical proofs concerning the theory of numbers proceed by a method called *reductio ad absurdum*. Proofs of this kind assume for the sake of argument that the contrary of a proposed theorem is true. It's then shown that such an assumption leads to a contradiction or an absurdity.

To illustrate the technique, let's follow Euclid's proof of the incommensurability of the side of a square with its diagonal. In modern terminology, we would set the sides of the square equal to 1 and state that the square root of 2, which is the diagonal, is an irrational number.

To prove the irrationality of $\sqrt{2}$ by *reductio ad absurdum* we begin by assuming that $\sqrt{2}$ is a rational number. If this assumption is true, a fraction can be found such that

$$\sqrt{2} = \frac{p}{q}$$

where p and q are integers. Let's assume, further, that the fraction

194

$\frac{p}{q}$ *is in its lowest terms.* Then at least one of the integers, either p or q, must be odd. It's clear that p cannot be odd since

$$\sqrt{2} = \frac{p}{q}$$

$$2 = \frac{p^2}{q^2}$$

$$2q^2 = p^2$$

The term $2q^2$ is obviously an even number so its equal p^2 must also be even. But if p^2 is even, p must also be even.

Since p is even, we can set $p = 2r$ where r is equal to half of p. Then, substituting in the the original equation, we get

$$2 = \frac{p^2}{q^2}$$

$$2 = \frac{(2r)^2}{q^2}$$

$$2q^2 = 4\,r^2$$

$$q^2 = 2r$$

By the same reasoning as before, q^2 and q must be even numbers. Our assumption that $\sqrt{2}$ is a rational number cannot be a valid one since it requires both p and q to be even, although the fraction $\frac{p}{q}$ was originally assumed to be in its lowest terms. The two requirements are obviously contradictory so the irrationality of $\sqrt{2}$ is proved by *reductio ad absurdum*. It follows, therefore, that the length of the diagonal of any square cannot be expressed as a ratio of natural integers.

This proof, elegantly simple as it is, was to have a profound effect upon Greek thinking. The Pythagorean theorem reads: *The sum of the squares built on the legs of any right triangle is equal to the square built on the hypotenuse.* The theorem so stated was known to be true for *all* right triangles. But many special cases of the theorem had been found in which all three sides were commensurable: 3:4:5, 5:12:13, etc. It was only natural to assume that

195

all such triangles have integral sides. The fact that more had not been found was not too surprising; after all, the ratios might consist of extremely large numbers—numbers that would tax to the limit the primitive calculating techniques of the Greeks. And so the matter lay, while the search went on for more and more Pythagorean triangles.

The Pythagorean theorem and the conjecture that all such triangles had sides that were commensurable led the Pythagoreans to conclude that *there was an inherent union between geometry and arithmetic.* The metaphysical implications of this conclusion only strengthened their belief in the philosophy that number rules the universe. Numerology was the balloon that supported their philosophy and the subsequent discovery by the Pythagoreans of irrational numbers was a dangerous and disruptive pin. Although they made a great attempt at secrecy, the truth became known, and the Pythagorean system of natural philosophy began to decline in importance. If number, after all, cannot account for the most significant branch of knowledge—geometry —how can it determine the course of the universe? And yet, even today many misguided individuals still cling to mystical beliefs in the supernatural power of numbers, little realizing that they worship an idol that was smashed 2,000 years ago.

How can Polaris, the North Star, be located?

If you're lost in the woods at night, finding the North Star may be a desirable accomplishment. To do so, first locate the constellation of Ursa Major, sometimes called the Great Bear or the Big Dipper. It consists of seven bright stars in the Northern sky and is visible from most of the Northern Hemisphere. The stars are arranged in the shape of a large pan or dipper, as shown in Figure 39. When you have found Ursa Major, draw an imaginary line through the two end stars of the Dipper bowl (usually called the Pointers), and this line will point almost directly in the direction of Polaris, the North Star. The distance between Polaris and the nearest Pointer is about five times the separation of the Pointers.

In looking for Ursa Major, it should be kept in mind that the constellation will not necessarily be in the position shown in the diagram when you find it. Just as with all the stars, the Big Dipper will rotate about the North Star during the course of the

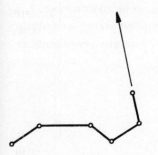

★ NORTH STAR

FIGURE 39. The pointers of the Big Dipper always point toward Polaris, the North Star.

night. The constellation will preserve its form, however, and the open portion of the dipper will always be in the general direction of the North Star.

Once you have found the North Star, it's a simple matter to find Ursa Minor, the Small Dipper, because Polaris is at the end of its handle.

How is the difference between two squares calculated?

There are many unexpected relationships associated with the differences between square numbers. Consider the squares of successive integers for example:

$$1^2 - 0 = 1 - 0 = 1 \quad \text{(Let's not slight zero!)}$$
$$2^2 - 1^2 = 4 - 1 = 3$$
$$3^2 - 2^2 = 9 - 4 = 5$$
$$4^2 - 3^2 = 16 - 9 = 7$$
$$5^2 - 4^2 = 25 - 16 = 9$$
$$6^2 - 5^2 = 36 - 25 = 11$$
$$7^2 - 6^2 = 49 - 36 = 13, \text{etc.}$$

197

It would seem from the calculations that the differences between the squares of successive numbers may be the infinite series of odd numbers (1, 3, 5, 7, . . .). Such a conjecture can be proved true, and a by-product of the proof will provide a way of calculating the difference simply and easily.

The method of proof involves using the letter n to denote the smaller of the two numbers. The larger number, then, is $n + 1$, and the squares are n^2 and $(n + 1)^2$, respectively. The difference that concerns us is $(n + 1)^2 - n^2$ where n can be any number, 1, 2, 3, . . .). In order to simplify the expression given above, let's square the term $(n + 1)$ just as we would 28, or 73, or any other number.

$$
\begin{array}{r}
n + 1 \\
\times\ n + 1 \\
\hline
n + 1 \\
n^2 +\ n \\
\hline
n^2 + 2n + 1
\end{array}
$$

Subtracting n^2 from $n^2 + 2n + 1$ leaves $2n + 1$. Put in the form of an identity, this expression becomes

$$(n + 1)^2 - n^2 = 2n + 1$$

Now let's examine the right-hand side of the expression $2n + 1$. If n is *any number*, it can be either odd or even. But whichever it may be, $2n$ will have to be even since the factor 2 is always present. If $2n$ must be even, then $2n + 1$ must be odd since the odd and even numbers alternate. Substituting $n = (1, 2, 3, . . .)$, the expression $2n + 1$ generates the series of odd numbers.

If it's desired to find the difference between 58^2 and 59^2, merely substitute $n = 58$ in the right-hand term of the formula

$$
\begin{aligned}
59^2 - 58^2 &= (2)(58) + 1 \\
&= 116 + 1 \\
&= 117
\end{aligned}
$$

and the difference is found to be 117.

Carrying this reasoning a step further, let's examine the differences between the squares of two numbers whose difference is 2.

$$2^2 - 0 = 4 - 0 = 4$$
$$3^2 - 1^2 = 9 - 1 = 8$$
$$4^2 - 2^2 = 16 - 4 = 12$$
$$5^2 - 3^2 = 25 - 9 = 16$$
$$6^2 - 4^2 = 36 - 16 = 20$$
$$7^2 - 5^2 = 49 - 25 = 24$$

This procedure also seems to have developed a recognizable series, the series of alternate even numbers. Using the same principles as before, the difference between the squares is

$$(n + 2)^2 - n^2 =$$
$$n^2 + 4n + 4 - n^2 =$$
$$4n + 4 =$$
$$4(n + 1)$$

The desired formula, then, is

$$(n + 2)^2 - n^2 = 4(n + 1)$$

If two squares are 108^2 and 106^2, their difference is $(4)(107) = 428$.

Since square numbers exhibit such exemplary behavior, perhaps a *general* formula can be developed which will simplify the calculation of the difference between *any two squares,* whatever they may be. Let n^2 be any square and let $(n + k)^2$ be any other square greater than the first. The letter k, of course, stands for the difference between the two numbers $n + k$ and n.

$$(n + k)^2 - n^2 =$$
$$n^2 + 2kn + k^2 - n^2 =$$
$$2kn + k^2 =$$
$$k(2n + k)$$

The general formula is then

$$(n + k)^2 - n^2 = k(2n + k)$$

Since the formula developed above is general in nature, it should be possible to derive the simpler formulas from it. This can be done by substituting $k = 1$ and $k = 2$ and simplifying. If $k = 1$, the right side becomes $2n + 1$; and if $k = 2$, it becomes $4(n + 1)$ —precisely the expressions that were obtained earlier.

Suppose the difference $28^2 - 23^2$ is to be calculated. Then $k = 5$ and $n = 23$; so $k(2n + k) = 5(46 + 5) = (5)(51)$, and the difference is 255.

It may be argued that mental gymnastics of this kind are of little practical value. After all, there are slide rules, tables of squares, books of logarithms, and buildings full of electronic computers to perform our most difficult calculations. And yet, one never ceases to be amazed at the beautiful and unexpected relationships hidden in our natural numbers. Without a knowledge of those relationships none of our modern wonders would have been possible.

How are stars used in celestial navigation?

At any instant, each heavenly body is at the zenith of a particular point on the earth's surface. This point which is directly under the heavenly body is called the *geographic position* of the body. It is the point of intersection with the earth's surface of an imaginary line drawn between the center of the earth and the body. The latitude and longitude of the geographic positions of many stars have been determined and are available in the *Nautical Almanac*. The only additional information needed to determine the position of the observer are the angles above the horizon of two or more stars.

Through the use of a sextant, it's possible to measure the *altitude* of a star at the place of observation. This is the angle between the star and the horizon. Subtracting this angle from 90° gives the angle between the star and the observer's zenith, or the *zenith distance* of the star. As you can see from Figure 40, the zenith distance is also the angle at the center of the earth between the observer and the geographic position of the star. Knowing this angle makes it possible to determine the distance between the observer and the geographic position of the star. A *nautical mile* is the length along the surface of the earth of one minute of arc measured on a great circle. Because of the oblateness of the earth, this distance varies from 6,046 feet at the equator to 6,108 feet at the poles. For practical work, however, a nautical mile is taken as 6,080 feet. The

zenith distance of a star, measured in minutes of arc, is then equal
to the distance in nautical miles between the observer and the geographic position of the star.

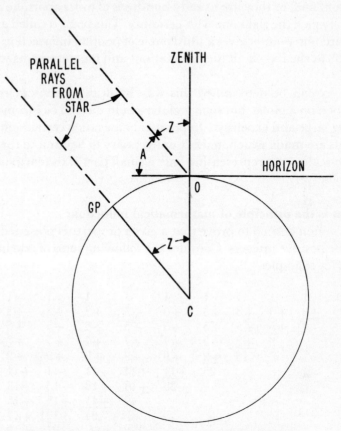

FIGURE 40. Using the stars for celestial navigation.

   With this information, a navigator knows how far he is from the
geographic position of the star, but he does not know his direction
from that point. In other words, he is somewhere on a circle whose
center is at the star's geographic position, and whose radius is
equal to the distance determined above. This circle is called a *circle of position.*

Observation of another star displaced somewhat from the first determines a second circle of position which intersects the first at two points. The observer must be at one of these points of intersection, and, as they are usually hundreds of miles apart, he can usually pick the right one with certainty. This point is called a *fix*. A third observation gives a third circle of position intersecting the others at the fix—if all the calculations and measurements went well.

A fix can be determined this way by drawing the circles of position on a globe, but such a globe would have to be inconveniently large and expensive. In practical navigation certain refinements are made which make it unnecessary to have more than a large-scale chart representing only a small part of the earth near the observer.

What is the principle of mathematical induction?

It's often desired to prove that a given property is possessed by all the positive integers. Consider the following sums of odd integers, for example.

1	1	1	1	1	1	1	1	1	1
+3	+3	+3	+3	+3	+3	+3	+3	+3	+3
4	+5	+5	+5	+5	+5	+5	+5	+5	+5
	9	+7	+7	+7	+7	+7	+7	+7	+7
		16	+9	+9	+9	+9	+9	+9	+9
			25	+11	+11	+11	+11	+11	+11
				36	+13	+13	+13	+13	+13
					49	+15	+15	+15	+15
						64	+17	+17	+17
							81	+19	+19
									100

It's apparent, as far as the additions were carried, that the sum of the consecutive odd integers, beginning with 1, is always a perfect square. And further, the sum is always equal to n^2 where n is the number of integers involved. If we add $1 + 3 + 5$, for example, $n = 3$ and $n^2 = 9$. Having noticed such an unexpected property of numbers, it's only natural to wonder whether it's true of *all* integers—or whether the series fails for some higher number of odd integers. The sums performed above could be continued for a lifetime or a hundred lifetimes but no matter how far they were carried, they could never prove the principle for *all* integers. One of the methods that mathematicians *do* use, however, is called *mathematical induction.** This method is a logical analogy to the game of standing a row of dominoes on end so that when the first is knocked over, it will cause the next to fall and so on until the entire row falls over. Mathematical induction uses the principle that if

a. A certain property is possessed by one particular integer, say 1 or 2, and

b. you can prove that whenever the property is possessed by any integer, *n*, it *must also be possessed by the next higher integer, n + 1*, then it follows that

c. the property is possessed by all integers greater than the one used in (*a*) above.

To apply mathematical induction to the principle of the odd integers, it must be restated in mathematical form as follows:

$$1 + 3 + 5 + \cdots + (2n - 3) + (2n - 1) = n^2$$

where $(2n - 1)$ is the *n*th odd integer and $(2n - 3)$ is the $(n - 1)$th odd integer. [Similarly, $(2n - 5)$ would be the $(n - 2)$th odd integer and so on.]

The first step in the proof is to show that the equation is true for a particular value of *n*. Let's pick *n* equal to 3. The theorem then states that

$$1 + 3 + 5 = 3^2$$

which is quite true since both sides of the equation are equal to 9. The theorem is true, therefore, for $n = 3$, and the first part of the proof is accomplished.

We must now prove that whenever the theorem is true for *any* number of terms, *n*, it is also true for the next higher number, $n + 1$. This will be accomplished if, after adding the $(n + 1)$th

* A rather well-used misnomer; although induction is a permissible scientific tool, it is never acceptable in a mathematical proof.

term to both sides of the equation, it can be reduced to the same form as the original. Let's now add the $(n + 1)$th term to both sides of the equation. This term is obtained by adding 2 to the last term of the series giving $(2n + 1)$.

$$1 + 3 + 5 + \cdots + (2n - 3) + (2n - 1) + (2n + 1) = n^2 + (2n + 1)$$

The right side of the equation can be factored to give $(n + 1) \times (n + 1)$ or $(n + 1)^2$ which is equal to the square of $(n + 1)$ by definition. The left side of the equation is merely the sum of the odd integers from 1 to the $(n + 1)$th odd integer. We have shown, therefore, that if the theorem is true for n terms it is also true for $n + 1$ terms. That completes the proof and the theorem must be true for *all* odd integers.

The remainder of the argument is as follows. In part *(a)* it was proved that the theorem was true for the first three odd integers, or $n = 3$. Part *(b)* tells us that if the theorem is true for $n = 3$, then it is also true for $n = 4$. But if it's true for $n = 4$, it is also true for $n = 5$, or $n = 6$, and so on indefinitely. It must be true, then, for all odd integers beyond the third. In practice $n = 1$ is generally chosen as the n of part *(a)*.

Mathematical induction can be used to prove other relations between the integers. A few are given below for the reader who cares to test his skill.

$$1 + 2 + 3 + \cdots + (n - 1) + n = \frac{n(n + 1)}{2}$$
$$2 + 2^1 + 2^2 + 2^3 + \cdots + 2^{n-2} + 2^{n-1} = 2^n - 1$$
$$1^2 + 2^2 + 3^2 + 4^2 + \cdots + (n - 1)^2 + n^2 = \frac{n(n + 1)(2n + 1)}{6}$$

Why can we see farther into a shower of rain than into a fog bank?

It's remarkable how much farther it's possible to see into a rain shower than into the cloud or bank of mist from which it falls. The reason behind this difference in visibility can be understood with the help of the following analysis. Though admittedly rough, the arguments should serve to illustrate the physical principles involved.

First of all, it must be pointed out that water vapor is an invisible gas and cannot, therefore, be a factor in this situation. Fog or mist consists of minute droplets of water suspended in air. When they become large enough they fall to earth as rain. Water droplets in fog or mist have a diameter of the order of 0.0004 inch while raindrops are about 0.02 inch in diameter.

Let's denote by V the volume of water (small or large drops) present in a unit volume of air. Now imagine that V is divided into drops of diameter d. The volume of each spherical drop will be $\dfrac{\pi d^3}{6}$; so the number of drops per unit volume is

$$N = V \div \frac{\pi d^3}{6}$$

or

$$N = \frac{6V}{\pi d^3}$$

Each of these drops will block off an area equal to its own cross section, $\dfrac{\pi d^2}{4}$; so the whole area blocked by the drops is equal to that quantity times the number of drops.

$$\text{Area blocked off} = \frac{6V}{\pi d^3} \times \frac{\pi d^2}{4}$$

or

$$\frac{1.5V}{d}$$

Therefore, the smaller the drops, the larger the blocked-off area, and the less transparent their aggregate.

In the case of a heavy mist, V is of the order of 10^{-6}, or 1 part of water in 1 million parts of air by volume. Oddly enough, V is just about the same for pouring rain. Using this approximate figure, and the drop sizes given above, it's possible to estimate the range of visibility in fog and in rain. Suppose the unit volume of air is a cube 1 foot on a side. Imagine that you are looking through a number of such volumes stacked one behind the other directly in your line of vision. The amount of light blocked

205

off by the drops in those volumes will be equal to $\dfrac{1.5V}{d} \times s$, where

s is the number of volumes involved. Since each volume is a 1-foot cube, s is also equal to the number of linear feet through which you are looking. In order to block off 80 per cent of the light, for example, we must have

$$\frac{1.5Vs}{d} = 0.8$$

For raindrops, the formula gives $s = 890$ feet, and for mist $s = 18$ feet. These visibilities are of the right order of magnitude.

The great dependence of the range of visibility on the size of the waterdrops is evident from the example. But it often happens, during a particularly heavy downpour, that the visibility is considerably reduced near the ground. This results from the splashing of raindrops into finer droplets as they strike objects near the ground, and the explanation fits in well with our reasoning.

Why is Venus called the queen of the evening sky?

At intervals of about 19 months, Venus puts on a spectacular evening show. At its best, Venus appears about 15 times as bright as Sirius, the brightest star! At that time, it even casts a noticeable shadow. When its evening brightness comes in December, it is not unusual for hundreds of people to phone observatories and inquire whether the star of Bethlehem has reappeared.

Because of the unique periods of revolution of Venus and the earth, it's easy to predict the times at which Venus will be visible as a morning or evening star. The time required for a planet to make one revolution around the sun is called its *sidereal period*. The earth's sidereal period is 1 year (365.256 days) and that of Venus, 224.701 days. As with all the planets, the orbits of Venus and the earth lie almost in the same plane (Figure 41). The two can be thought of as runners who start off on a race when they line up with the sun. Venus, being closer to the sun, has the advantage of the inside track—a path much shorter than that of the earth. In addition, Venus travels at about 22 miles per sec-

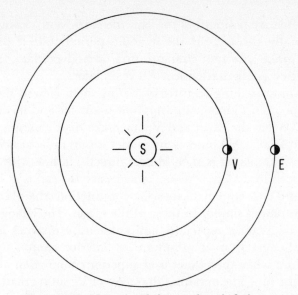

FIGURE 41. Venus and the earth at inferior conjunction.

ond compared to the earth's $18\frac{1}{2}$ miles per second, and so over-
takes the earth at regular intervals. The length of these intervals
is called the *synodic period.*

The sidereal period of Venus is about seven tenths of a day
longer than 32 weeks, while our sidereal period is about 1 day
more than 52 weeks. Expressed in terms of a year, the sidereal
period of Venus is about $\frac{32}{52}$ or $\frac{8}{13}$ of a year. If Venus makes one
trip around the sun in $\frac{8}{13}$ of a year, the number of trips it will
make in 1 year equals $1 \div \frac{8}{13} = \frac{13}{8} = 1\frac{5}{8}$ trips. During 1 sidereal
year, Venus gains $\frac{5}{8}$ of a lap on the earth. The number of years
it will take to gain a whole lap on the earth is $1 \div \frac{5}{8} = \frac{8}{5}$ or 1.6
years. This is the synodic period of Venus.*

* A more precise determination of the synodic period can be obtained by using the for-
mula $\frac{1}{S} = \frac{1}{P} - \frac{1}{E}$, where S is the synodic period, E is the sidereal period of the earth, and P
is the sidereal period of the planet. For planets beyond the earth's orbit, the formula be-
comes $\frac{1}{S} = \frac{1}{E} - \frac{1}{P}$. A more exact figure for the average synodic period of Venus is 583.9
days.

207

The relative positions of the sun, the earth, and Venus change continually throughout the synodic period; but if the three bodies make a certain configuration on a given date, they will repeat that configuration just 1.6 years later.

The most striking features of Venus when viewed through a telescope are its change in apparent diameter and its changing phases. When the earth and Venus make their closest approach —about 26,000,000 miles—the planet is said to be at *inferior conjunction*. The planet is then almost directly in line with the sun and is seen in the "new moon" stage since the sun is at its back. Prior and subsequent to inferior conjunction, the planet goes through phases similar to those of the moon. The change in apparent diameter is especially noticeable as its distance from the earth changes from a minimum at inferior conjunction to a maximum when the planet is at superior conjunction at the far side of the sun. The apparent diameter of Venus is greatest when it is closest to the earth, but it is then in the "new" phase and very little of its visible surface is illuminated by the sun. On the other hand, its apparent diameter is only one-sixth as great at the "full" phase since at that time it is at its greatest distance from the earth. These effects combine to produce greatest brilliance about one month before and after inferior conjunction when about one-quarter of her earthward hemisphere is illuminated.

During the months prior to inferior conjunction, Venus is seen as an evening star in the western sky. Following inferior conjunction, the planet changes to a morning star seen before dawn in the eastern sky. At times, Venus is visible from points in the United States for more than three hours after sunset or before sunrise. Telescopic observation of the planet is best carried on in the daytime, however, because at night, Venus is too low in the sky for "good seeing."

Venus was at inferior conjunction on August 31, 1959, and will be again on April 6, 1961. Subsequent recurrences of inferior conjunction can be approximated closely by adding 1.6 years (or 583.9 days) to that date. Venus will then be at her brightest as an evening star about one month prior to inferior conjunction.

Mercury, the other planet whose orbit lies within that of the earth, is closest to the sun and so has the swiftest motion of all. At its brightest, Mercury appears to the unaided eye as a star nearly as bright as Sirius but so near the sun that it's never seen at night from northern latitudes. For this reason, Mercury is never too conspicuous in the sky.

Mercury is visible to the unaided eye only during those intervals when it is far enough away from the sun in the sky. These intervals are about two weeks long and there are usually six of them each year. The intervals are centered about the dates on which Mercury is at its greatest angular distance from the sun —an angle that may vary from 18 to 28 degrees from one synodic period to the next.

Mercury takes 116 days on the average to gain a lap on the earth, so its synodic period is 116 days or nearly 4 months long. Mercury is visible twice during each synodic period. As with Venus, Mercury is an evening star prior to inferior conjunction and a morning star following inferior conjunction. It is best seen as an evening star in the spring or a morning star in the autumn. At these times an imaginary line drawn from Mercury to the setting or rising sun is most nearly vertical in the sky. This gives the planet the maximum height above the horizon where it can be seen at dusk (in the spring) and dawn (in the autumn).

What are prime pairs?

A glance at a list of prime numbers shows that they often go in pairs. A few examples of prime pairs are 11, 13; 17, 19; 59, 61. Each pair of primes is separated only by the even number that would ordinarily fit between them. No end has been found to the succession of prime pairs; so mathematicians wonder whether such pairs, like the primes themselves, go on forever. A great deal of effort has gone into the study of prime pairs in modern times, but no light has yet been shed on the subject. It's not known how to prove or disprove the conjecture and the problem remains unsolved. Any ideas?

What is the magnitude of a star?

For thousands of years man has gazed inquiringly toward the stars in an effort to understand and explain them. Little was accomplished, however, until about two thousand years ago, when Ptolemy noticed that the brightness of the stars varies considerably from one to the next. Although he had no instruments with which to measure light intensity, he decided to classify the stars by their brightness. With no more than his own eyes to rely upon, he divided the stars into six classes, or *magnitudes*, in reverse order of their brightness. The higher the magnitude, the fainter the star.

Over the intervening centuries, Ptolemy's art developed into the exact science of astronomy with its fabulous instrumentation, but the old concept of magnitudes persists to this day. With the advent of adequate measuring devices, astronomers measured the brightness of Ptolemy's stars and found each magnitude to be about $2\frac{1}{2}$ times as bright as its higher numbered predecessor. Take any two stars, for example. If one is of the fourth magnitude, and the other of the fifth, the first will be about $2\frac{1}{2}$ times as bright as the second. More precisely, the relation between Ptolemy's magnitudes is:

Magnitude	Relative brightness
6.0	1.000
5.0	2.512
4.0	$(2.512)^2$
3.0	$(2.512)^3$
2.0	$(2.512)^4$
1.0	$(2.512)^5$

Soon after Galileo's perfection of the telescope, countless new and faint stars (or more properly, nebulae) were discovered and it was necessary to extend the number of magnitudes to include them. This was done merely by applying the factor of 2.512 each time a new magnitude was needed.

210

Going in the other direction, it was later discovered that Vega is about $2\frac{1}{2}$ times as bright as a standard first magnitude star so it became a star of 0 magnitude. Similarly, Canopus is almost $2\frac{1}{2}$ times as bright as Vega, so it has a magnitude of -0.9.

The magnitudes of the stars, then, are measures of their brightness. They are expressed by numbers that get larger as the stars are fainter. Since these numbers express the brightness of stars *as we see them,* they are usually called *apparent* or *visual magnitudes.* But the amount of light that reaches us from a star also depends on the star's distance from us (and of course on any intervening matter that might dim its light). If these factors are taken into account, it's possible to determine the *absolute magnitude* of a star. This is the magnitude that a star would have if it were located at a standard distance from the earth, with no intervening obstructions. Astronomers have arbitrarily fixed this standard distance at $32\frac{1}{2}$ light years, or 10 parsecs. The sun has an apparent magnitude of 26.7 because it is so close to us. When placed at the standard distance, its absolute magnitude becomes $+5$—the brightness of a star faintly visible to the unaided eye. Apparent magnitudes can be changed into absolute magnitudes by using the relation:

$$M = 5 + m - 5 \log d$$

where M is the absolute magnitude, m is the apparent magnitude, and d is the distance in parsecs.*

What is the principle of least time?

Have you ever searched the sky fruitlessly for a high-speed aircraft even though its engines are clearly audible? When you do find a plane in the sky, it usually turns up at the "wrong" place—a place quite removed from the apparent source of the

* Using the definition of magnitudes given earlier, the apparent brightness of a star is $(2.512)^{6-m}$ and the absolute brightness is $(2.512)^{6-M}$. These brightnesses are inversely proportional to the square of the distances from earth, so the proportion becomes

$$\frac{(2.512)^{6-M}}{(2.512)^{6-m}} = \frac{d^2}{10^2}$$

Solving this equation gives the desired relation.

engine noise. One of the major reasons for this kind of auditory illusion is the *principle of least time.*

The principle in its usual form tells us that a sound will travel from source to observer by the path that takes the least time. That statement is so obvious as to be true by definition; somewhat less obvious, however, is the inescapable implication that sound does not necessarily travel by the shortest path.

The effect was probably first noticed in connection with shells whistling overhead. In modern times, we find a similar lack of correlation between the actual direction of a plane or missile and the direction from which its sound seems to come.

There is a simple mathematical relationship between the trajectory of a supersonic device and the direction from which its sound first reaches an observer. Imagine a high-speed jet plane traveling from point P to Q, as in the upper drawing in Figure

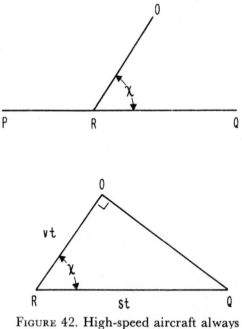

FIGURE 42. High-speed aircraft always turn up at the "wrong" place.

212

42. An observer is located at point O. If the plane is moving at a supersonic speed s, and the velocity of sound is v, then the sound will first reach the observer by a path RO such that

$$\cos x = \frac{v}{s}$$

The sound will first reach the observer by a path that makes an angle x to the trajectory.

From the lower drawing in the diagram, it is possible to deduce the fact that our sense of direction of the sound will always be wrong by 90°. The distance, velocity, and time of a moving object (or sound) are related by the well-known expression:

$$\text{Distance} = \text{velocity} \times \text{time}$$

or, in algebraic shorthand,

$$\text{Distance} = vt$$

If it takes t seconds for the sound to move from R to O, the distance $RO = vt$. During this length of time, the moving plane will have traversed a greater distance $= st$. But

$$\cos x = \frac{\text{side adjacent}}{\text{hypotenuse}} = \frac{RO}{RQ} = \frac{vt}{st} = \frac{v}{s}$$

It follows that RO and RQ are a side and hypotenuse, respectively, of a right triangle, and angle $O = 90°$. When the sound first reaches O, the plane will have reached the point Q, precisely 90° removed from the direction of its sound. The 90° error is true for *any* supersonic speed the plane may have. At subsonic speeds, the effect is still noticeable but less than 90° in magnitude.

What is the Great Theorem of Fermat?

The Great Theorem of Fermat is an excellent example of a simply stated problem that has baffled the greatest mathematical minds for over three hundred years. The roots of the problem go back to ancient Egypt where every self-respecting carpenter knew that a triangle having sides in the ratio of 3:4:5 includes a right angle. During the third century, Diophantus of Alexandria was struck by the mathematical beauty of the 3:4:5 triangle and he began to speculate on whether any other right triangles had sides that were measured in integers. Diophantus knew that any right triangle obeyed the Pythagorean theorem,* so the problem reduced to that of finding two integers, the sum of whose squares would equal the square of a third. If a, b, and c are the sides of a right triangle, the problem is to find whole-number solutions to the equation $a^2 + b^2 = c^2$. Right-angled triangles of this kind are now known as Pythagorean triangles.

In the year 1621, Pierre de Fermat obtained a copy of the new French translation of Diophantus' book, *Arithmetica*. In this book, Diophantus not only showed that there were other Pythagorean triangles, but that there were an infinite number of them. To find such triangles, take any two numbers x and y such that $2xy$ is a perfect square ($x = 2$ and $y = 4$, for example). Then $a = x + \sqrt{2xy}$; $b = y + \sqrt{2xy}$; and $c = x + y + \sqrt{2xy}$. In the example mentioned above,

$$a = 2 + \sqrt{2(2)(4)} = 6$$
$$b = 4 + \sqrt{2(2)(4)} = 8$$
$$c = 2 + 4 + \sqrt{2(2)(4)} = 10.$$

When Fermat read Diophantus' discussion of Pythagorean triangles, he made a short note in the margin to the effect that whereas the equation $a^2 + b^2 = c^2$ has an infinite number of integer solutions, *any equation* of the type $a^n + b^n = c^n$ where n is larger than 2 has *no* solution whatsoever. "I have discovered a truly wonderful proof of this," wrote Fermat, "which, however, this margin is too narrow to hold."

For 300 years, the world's best mathematicians have tried to discover the proof of Fermat's theorem. Although considerable progress has been made during that length of time, no such proof has yet been found. Euler demonstrated the impossibility of integer solutions of the equations where n is equal to 3 and 4,

---

* The square erected on the hypotenuse is equal to the sum of the squares erected on the other two sides.

and Dirichlet did the same for the equation of the fifth power. We now know that no solutions exist for any values of the exponent *n* smaller than 269, but no general proof, good for *any* value of *n* has ever been constructed.

It's possible that even a mathematician of the stature of Pierre de Fermat could have made a mistake in his proof, but the search still goes on. Fermat made many remarks in letters and marginal notes concerning new proofs that he had discovered but he seldom published them. He seemed more interested in discovering proofs than in publicizing them. In almost every case, however, when Fermat said he had a proof, one was later found. For that reason, if for no other, mathematicians will undoubtedly search for the proof of the Great Theorem, or for an example that will disprove the theorem. But since such an example would involve numbers with exponents larger than 269, the latter course will be a long and arduous one.

What is the weight of the atmosphere?

Although the pressure of the atmosphere varies from place to place and even from time to time at the same place, the variations are small and the *standard atmosphere* is usually taken to be 14.7 pounds per square inch. With this assumption in hand, it becomes possible to estimate the weight of the atmosphere. Imagine a square horizontal surface, 1 inch long on each side. Now imagine a tall column of air having a cross section of 1 square inch and extending from the surface to the uppermost limits of the atmosphere. The weight of that column of air is just 14.7 pounds. Multiplying this number by the earth's area in square inches will give a fair estimate of the atmosphere's weight.

The area of a sphere is $4\pi R^2$; so the approximate area of the earth is

$$4\pi(3{,}960)^2 \,(5{,}280)^2 \,(12)^2 = 7.9 \times 10^{17} \text{ square inches}$$

The weight of the atmosphere is then about $14.7 \times 7.9 \times 10^{17} = 10^{19}$ pounds, approximately. This amounts to 10 billion billion pounds. Since the mass of the earth is about 10^{25} pounds, the atmosphere weighs about one-millionth as much as the earth—a not insignificant amount.

How many stars can a telescope "see"?

Have you ever glanced skyward on a clear moonless night and remarked on the number of stars? At first glance they seem innumerable. And yet if your attention is concentrated on a limited area, such as the bowl of the Big Dipper, the stars in that area can easily be counted. In fact, if you were to count all the stars that you can see with the naked eye, the number would amount to only about 2,500 or 3,000. If you double that number to account for the unseen hemisphere of the heavens, the total would reach only 5,000 or 6,000.

These have been classified into six *magnitudes* in order of brightness. By following a similar procedure with powerful telescopes, astronomers have counted the stars down to the twentieth magnitude of brightness. The table below gives the number of stars brighter than a given magnitude down to the twentieth.

Number of stars brighter than a given magnitude

MAGNITUDE	NUMBER	RATIO
2	40	
3	135	3.4
4	450	3.3
5	1,500	3.3
6	4,800	3.2
7	14,300	3.0
8	41,300	2.9
9	117,000	2.8
10	324,000	2.8
11	868,000	2.7
12	2,260,000	2.6
13	5,700,000	2.5
14	13,800,000	2.4
15	32,000,000	2.3
16	70,800,000	2.2
17	149,000,000	2.1
18	296,000,000	2.0
19	560,000,000	1.9
20	1,000,000,000	1.8

There are 135 stars brighter than the third magnitude, 5,700,-000 brighter than the thirteenth, and 1,000,000,000 brighter than the twentieth! The number in the third column is the ratio of each star count to its predecessor, and will be valuable later in estimating the total number of stars.

If you were to look through a telescope having a 1-inch aperture, or objective lens diameter, you would be able to discern stars as faint as the ninth magnitude. In general, the relationship between the aperture of a telescope and the faintest star it makes visible* is given by:

$$m = 9 + 5 \log a$$

Where m is the star's magnitude and a is the aperture of the telescope. The limiting magnitudes that can be seen with telescopes of various size are given in the following table.

Aperture, inches	Limiting magnitude	Aperture, inches	Limiting magnitude
1	9.0	24	16.0
2	10.5	36	16.8
3	11.4	40	17.0
4	12.0	60	17.9
5	12.5	100	19.0
6	12.9	200	20.5
10	14.0	400	22.0
12	14.4	800	23.5
20	15.5	1,000	24.0

The 6-inch telescope, popular with amateurs, makes visible stars of magnitude 12.9, or, in round numbers, 13. This corresponds to 5,700,000 stars as seen in the table given earlier. This is over 1,000 times as many stars as are visible to the unaided eye! The 200-inch Hale reflector on Mount Palomar makes visible 1,000,000,000 stars, or 1,000,000 times the number that can be seen without a telescope.

Somewhat fainter stars can be photographed with the same

* For a derivation see: John C. Duncan, *Astronomy*, 5th ed. (New York: Harper & Brothers, 1954), p. 338.

217

instruments. The 200-inch Hale telescope has photographed stars as faint as the twenty-third magnitude.

By referring to the first table we can deduce the total number of stars in the galaxy. The figures in the third column, you will recall, give the ratio of each number of stars to its predecessor. Although the number of stars increases with each magnitude, the ratios diminish at a constant rate and would become unity at about the twenty-eighth magnitude. This would mean that the number of stars brighter than the twenty-eighth magnitude is the same as the total number brighter than the twenty-seventh. For this to be true, there must be no stars in the twenty-eighth magnitude. Extrapolations of this kind give answers between 10 billion and 30 billion stars as the total number in the galaxy. There is, of course, a great deal of uncertainty in these estimates. If correct, they are based on from 10 to 30 invisible stars for each star that can be seen with the world's most powerful telescopes. In addition, there is a great deal of obscuring matter in the galaxy that must hide great numbers of stars from view. And the theory that the ratios will continue to perform for future magnitudes as they have in the past may be a good *guess*, but it's hardly a proven *fact*. For these reasons, it would seem that the estimate of a few tens of billions of stars must be *far too low*.

It would require a telescope over 80 feet in diameter to count the stars to the twenty-seventh magnitude—at least from the surface of the earth. On the moon, however, with continual "perfect seeing" and no shimmering atmosphere, it should be possible to photograph distant stars with relatively small telescopes. For this reason, the next great breakthrough in astronomy will probably follow the establishment of an observatory on the moon.

Why do telescopes have such large diameters?

If you are thinking in terms of increased magnification, you will have to accept a very large and probably surprising demerit! The magnification of a telescope depends only on its length—not on its diameter. Then why do astronomers struggle to build and

218

manipulate such mammoth telescopic monsters as the 200-inch reflector on Mount Palomar? Imagine pouring a 20-ton piece of glass followed by the patient business of grinding and polishing to dimensions closer than one hundred-thousandth of a millimeter! A huge 500-ton mounting is required to hold the giant telescope and an electronic brain must be used to regulate its movements. In all, 20 years passed from inception of the idea by George Ellery Hale to completion of the project in 1948.

To fully appreciate the need for large-diameter telescopes we must go back to the latter part of the eighteenth century, and witness the greatest sensation since the days of Galileo. An obscure, amateur astronomer had discovered, of all things, a new planet—Uranus! William Herschel had completely upset the entire astronomical world. No one was expecting any exciting surprises—especially not from astronomers. They had become the dull bureaucratic timekeepers of the universe, duly recording their ever-more-precise measurements. Astronomy seemed to have reached its goal, and then, out of nowhere, Herschel opened up fantastic new goals with his amazing discovery. He did all this with a new kind of telescope.

The mirror (or *objective lens*) at the lower end of Herschel's reflecting telescope was twice the size of any known telescope. It was so unusually wide that a man could put his head into the tube. Why this abnormal width when magnification depends only on length? Herschel's answer was simplicity itself—to gather light! He wanted to see stars whose light was feeble, stars and nebulae more distant than had ever been seen before. It was this principle, he explained, that had made Uranus visible to him. A telescope shows more stars not because of its greater magnification, but because of its greater light-gathering ability. Like an eye with a larger pupil, it concentrates more rays of light from a star into a single image. In modern photography (unknown to Herschel), we accomplish the same purpose by "opening up" the lens of a camera in dim light.

The light-gathering power of a telescope is defined as the ratio of the amount of light from a given star entering the *objective* (or large lens of the telescope) to the amount entering the pupil of the unaided eye. The light-gathering power is then,

$$L = \frac{\text{area of objective}}{\text{area of pupil}} = \frac{\pi a^2}{\pi b^2} = \frac{a^2}{b^2}$$

where a is the diameter of the objective and b is the maximum diameter of the pupil. If we let $b = \frac{1}{3}$ inch, the light-gathering power of a telescope of a inches *aperture* becomes

$$L = 9a^2$$

The study of optics shows that the maximum useful magnification of a telescope is given by $M = 4a$. Even at this magnification, stars are at such prodigious distances that they appear only as points of light. Since the image of a star is too small to be seen as a disk, the effective size of all star images on the retina is the same. For this reason, the apparent brightness of a star depends only on the light-gathering power of the telescope. A 2-inch pair of binoculars makes a star appear $9 \times 2^2 = 36$ times brighter than to the unaided eye. Similarly, a 6-inch telescope increases the brightness 324 times, a 10-inch telescope 900 times, and a 200-inch telescope 360,000 times. This principle enables large telescopes to see millions of faint stellar objects that would otherwise be invisible.

Having made this discovery, Herschel, who was perhaps the greatest of all astronomers, went on to investigate a brilliant inspiration. He decided to measure the "space-penetrating power" of the eye and of his telescopes! When he announced his objective, a critic accused Herschel of acute mental deficiency, brought on, no doubt, by advancing senility; he indicated that such a boast would be more appropriate to a poet than to an astronomer. But Sir William was to attain his end and become the first astronomer to make reasonably accurate measurements of linear distances into space.

His ruler was the light-year, the distance traveled by light in the course of a year—approximately 6 million million miles. He reasoned that if we could only learn the distance of a single star, the distances of all the others might be calculated from it. After several vain attempts at this first measurement he concluded that the nearest stars, such as Sirius, must be at least 3 light-years away. On the basis of this number he proceeded to lay his yardstick among the stars.

Here is the reasoning he used. Since ancient times, all the stars visible to the eye had been divided into six classes or *magnitudes* in order of brightness. From his own visual observations and the examination of much data he became convinced that each magnitude was about half as bright as the preceding one. More precisely, he found that a difference of five classes or magnitudes corresponds to a brightness ratio of 100. This means that each magnitude is $\frac{1}{2.512}$ times as bright as the preceding one.* Stars of the seventh magnitude (just beyond the range of human vision) would then be $\left(\dfrac{1}{2.512}\right)^6 = 0.0041$ times as bright as first-magnitude stars. But brightness is inversely proportional to the square of the distance so a seventh-magnitude star must be $1/\sqrt{0.0041} = 15.6$ times as far away as a first-magnitude star.

If we take Herschel's guess of 3 light-years for Sirius (the true distance is 8.6 light-years), the distance to seventh-magnitude stars must be 15.6 × 3, or about 47 light-years away. So the "penetration" of the human eye into space must be of the order of a few tens of light-years, according to Herschel's reasoning. This was the extent of the "universe" before Galileo pointed his first telescope toward the sky. Now Herschel had the information necessary to determine the penetration of his telescopes into space.

A 12-inch mirror increases the brightness of a star by an amount equal to its light-gathering ability,

$$L = 9a^2 = (9)(12)^2 = 1{,}296$$

or 1,296 times. This corresponds to a distance ratio of $\sqrt{1{,}296} = 36$, so a 12-inch telescope should penetrate 36 times deeper into space than the human eye; a 48-inch telescope, 45 times deeper.

---

* For a more complex analysis, see the discussion of star magnitudes, p. 210.

From such considerations Herschel estimated the galaxy to be 8,000 light-years in diameter and 1,600 light-years thick. He had measured our island universe. According to present estimates, the galaxy is about 100,000 light-years in diameter. But this difference does not detract from Herschel's greatness in having been the first man to penetrate into infinity! He showed astronomers the way into space—it hardly matters that his yardstick was a bit too long.

What is the sieve of Eratosthenes?

Whenever we multiply several numbers together, we get another; for example, $2 \times 3 \times 5 = 30$. If a number such as 30 can be expressed as a product of smaller numbers, the latter are called *factors*. But not every number is factorable; 31 cannot be expressed as a product except for the uninteresting case of $1 \times 31 = 31$. Numbers such as 31 which are not factorable are called *prime numbers*.

An important property of the primes, which may seem obvious, is that every number can be expressed as a product of prime factors. In the example mentioned above, 2, 3, and 5 are prime numbers and any number of additional examples can be found. Somewhat less obvious, however, is the fact that every number can be expressed as a product of prime factors in *only one way*. The only set of prime factors that can produce 30 are 2, 3, and 5. For some numbers a given prime factor may be repeated, but there is always one unique set of prime factors associated with each number.

Since every number can be made up of prime factors, we may want to know which numbers are prime—and how to find them. A simple method, known to the ancient Greeks, is called the sieve of Eratosthenes. Suppose, for example, that one wants to find all the prime numbers less than 100. He would proceed by writing all the numbers from 1 to 100. The first odd number, 1, is prime. The first even number, 2, is also prime, but no other even number can be prime; so we must cross out all the even numbers $(4, 6, 8, \ldots)$ from 4 to 100. The next number in the list is 3, which must be prime since it is not divisible by any smaller number. Now we must cross out all the multiples of 3 (6, 9, 12, ...). Some numbers, such as 6 and 12, will be crossed out twice, which is not unexpected, since many composite numbers will have at least two prime factors. In this way we proceed to the primes 5 and 7 and cross out their multiples. All the numbers from 1 to 10 are now crossed out, as are a great many between 10 and 100. The numbers that remain are all the primes between 1 and 100.

Perhaps you are wondering how we can be sure that the larger numbers are prime when we eliminated only the multiples of the first few prime numbers. The reasoning runs as follows. The number 100 can be factored in many ways but each of these eventually resolves to a unique set of primes; $100 = 25 \times 4 = 5 \times 5 \times 2 \times 2$; $100 = 50 \times 2 = 25 \times 2 \times 2 = 5 \times 5 \times 2 \times 2$; $100 = 10 \times 10 = 5 \times 2 \times 5 \times 2 = 5 \times 5 \times 2 \times 2$. All the prime factors of 100 are less than 10. Even if we select a number (less than 100 of course) that has only two prime factors, one of these must be less than 10. Stated another way, if a number has two prime factors larger than 10, the number must be larger than 100. So the sieve of Eratosthenes retains all the prime numbers from 1 to 100. This method can be used to find the primes up to any desired number merely by crossing out multiples of all primes up to the square root of the highest number in the sequence.

What is π?

The ancients knew that the ratio of the boundary of a circle to its diameter is the same for all circles. Today we call that ratio π because, if we try to express it as an arabic number, it stretches out forever, like $\sqrt{2}$. Since it can't be represented by a short number, it's more convenient to use the Greek letter until such time as a numerical value is needed for computations. Then we use its arabic equivalent which, correct to four decimal places, is 3.1416.

The Old Testament (2 Chron. 4:2) tells us that π is 3. "Also he made a molten sea of ten cubits from brim to brim, round in compass, and five cubits the height thereof; and a line of thirty cubits did compass it round about." So Solomon's molten sea had a circumference that was 3 times the diameter. The Hebrews, like the ancient Babylonians, were apparently content to use $\pi = 3$. In the not too distant past, a state legislature in this country had introduced a bill that would have restored π to its Biblical value. This, of course, was carrying things a bit too far, and the bill didn't pass.

By 1500 B.C., the Egyptians had arrived at a value (3.16) which is quite good and permits working to an accuracy of 1 per cent. Archimedes was satisfied with a value between $3\frac{1}{7}$ and $3\frac{10}{71}$. About A.D. 480, an irrigation engineer by the name of Tsu Ch'ung Chih came up with an astonishingly close estimate for his time. He found that π lies between 3.1415926 and 3.1415927. One thousand years later, the Arab Al Kashi determined $\pi = 3.1415926535897932$. By 1720, the Japanese Matsunaga had determined π correct to 50 decimal places. Today we know π correct to hundreds of decimal places.

While knowing π to such great accuracy may seem reassuring, it's really quite unnecessary for all practical purposes. Ten decimal places are enough to give the circumference of the earth correct to within a fraction of an inch. Even in the design of rocket engines, four decimal places are usually enough; π in the sky is apparently 3.1416.

What is the basis of arithmetic?

When all of us were very young we were taught the addition table; $1 + 1 = 2, 1 + 2 = 3, \ldots$ In the course of this instruction we discovered, or were told, that $2 + 3 = 3 + 2$. Similarly, $4 + 5 = 5 + 4$ and $2 + 5 = 5 + 2$. We were informed that such identities are not accidental but constitute a general rule. Mathematicians put it precisely by saying *the sum is independent of the order of the terms*. This principle, known as the *commutative law,* can be written in symbols,

$$a + b = b + a$$

We learned, in much the same way, that $(1 + 2) + 3 = 1 + (2 + 3)$; that is, it's immaterial in which order we add the terms, for we will get 6 in any event. So addition is also an *associative* process.

Take the sum of larger numbers, for example,

$$
\begin{array}{r}
21 \\
17 \\
\underline{33} \\
71
\end{array}
$$

If this procedure is analyzed we get

$$
\begin{aligned}
& 21 + 17 + 33 = \\
& (20 + 1) + (10 + 7) + (30 + 3) = \\
& (20 + 10 + 30) + (1 + 7 + 3) = \\
& (30 + 30) + (8 + 3) = \\
& 60 + 11 = 71
\end{aligned}
$$

The commutative and associative principles play an important role in these operations.

When multiplication and addition are jointly involved in an operation, another principle comes into play. We can obtain the product $5 \times (3 + 4)$ in two ways; $5 \times 7 = 35$, or $15 + 20 = 35$. Multiplication is said to be *distributive* with respect to addition. In symbolic form,

$$
a \times (b + c) = ab + ac
$$

We can think of this principle as allowing us to distribute the a across the terms within the parentheses.

The associative, commutative, and distributive properties of the operations are the basic principles upon which arithmetic is built.

Why is the color of incandescent light different from sunlight?

Energy reaches the earth from the sun by radiation; that is, by waves transmitted through the intervening space at the speed of light. These electromagnetic waves are entirely similar to those used in television broadcasting, although the latter are several million times greater in length.

225

Although light seems to travel instantaneously, we now know that its speed is finite—about 186,000 miles per second. (Galileo, one of the first truly great applied mathematicians, suspected this fact and suggested ways and means of proving it.) As it moves from one point to another, light seems to make use of a wavelike motion, much as water waves transfer mechanical energy from one place to another on the surface of the ocean. Since such waves are normally identical, or *periodic,* we can speak of a *wavelength* as the distance from one crest to the next (Figure 43). For the same

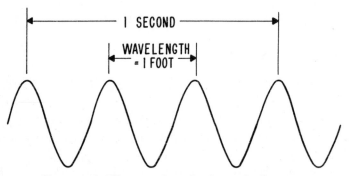

Figure 43. The wavelength of a periodic wave.

reason, light has, as one of its properties, a wavelength which depends upon the distance from one "crest" to the next. A little reflection will show that wavelength and the speed of light can be related by the simple expression:

$$\text{Speed } c = \text{wavelength } \lambda \times \text{frequency } f$$

where *frequency* is the number of vibrations per second of the wave.

The following table will give an idea of the approximate wavelengths used for various purposes:

Radio, television, communications .... 1,000 km to 1 cm
Infrared......................... 0.03 to 0.000076 cm
Visible light .... 0.000076 to 0.000040 cm
Ultraviolet .... 0.000040 to 0.0000013 cm
X rays .... 10^{-6} to 10^{-9} cm
Gamma rays .... 10^{-8} to 5×10^{-11} cm

It's evident that the energy reaching us from the sun covers an enormous range of wavelengths. By far the greatest amount, however, is concentrated in the region from perhaps 10^{-6} to 10^{-3} centimeter. This would include all the visible portion of the electromagnetic spectrum and most of the ultraviolet and infrared wavelengths. In the visible portion there is a definite relation between wavelength and our familiar colors of the rainbow.

Color	Wavelength, millionths of a centimeter
Violet	40–42
Blue	42–49
Green	49–57
Yellow	57–59
Orange	59–65
Red	65–76

Of these, the sun shows a marked preference for 49×10^{-6} centimeter, for at that wavelength solar radiation reaches its maximum intensity. It reaches a peak in the blue-green region and falls off rapidly at higher or lower wavelengths. Violet and red light are about 80 and 65 per cent as bright, respectively, as blue-green light.

So sunlight is a mixture of varying amounts of all colors of light. To duplicate the illuminating properties of sunlight, an artificial source must approximate that mixture—and to do that, the glowing filament within a lamp bulb must reach the effective temperature of the surface of the sun!

As any substance is heated, we find that it radiates energy. As we would expect, the amount of radiant energy goes up rapidly with increased temperature. If the substance gets hot enough, part of its radiated energy is in the visible region and it begins to glow a dull red. As the temperature is increased further, the red gets brighter and changes to yellow. White heat is hotter still. So increased temperature not only induces a change in the intensity of radiated energy—*it produces a change in the color of that radiation;* or more properly, in its wavelength. As the temperature of a substance is increased, the wavelength of maximum radiation is *displaced* toward the shorter wavelengths. A German physicist, Wilhelm Wien (1864–1928), formulated the simple law that covers this shift in wavelength. It's known as *Wien's displacement law* and states that the wavelength of maximum radiation, λ (lambda), is inversely proportional to the absolute temperature, *T*.

$$\lambda = \frac{0.2885}{T}$$

where λ is in centimeters and *T* is in degrees Kelvin (K) = degrees centigrade + 273. We have already noted that the wavelength of maximum solar radiation is 49×10^{-6}. The effective surface temperature of the sun can then be calculated by using Wien's formula.

$$T = \frac{0.2885}{49 \times 10^{-6}} = 5900° \text{ K}$$

To approximate the character of sunlight, a filament must reach a temperature of 5900° K. For practical reasons, incandescent lamps operate at considerably lower temperatures and most of their light is in the yellow and red portion of the spectrum.

The effective temperature of the sun's outer layers can be verified by use of the Stefan-Baltzmann law which states that the intensity of radiation is proportional to the fourth power of the temperature:

$$I = (1.36 \times 10^{-12})T^4$$

where *T* is in degrees Kelvin and *I* is the number of calories per square centimeter per second. Measurements of the *solar constant I* indicate that 1.94 calories is received above the atmosphere on a one square centimeter surface in one minute. But the intensity at the surface of the sun will be much greater—to find out *how much* greater, we will have to derive the *inverse-square law*.

Imagine a luminous body, the sun for example, at the center of a sphere of radius *R*. Let the sun's radius be *r*. The intensity of the radiated energy at the surface of the sun will be

228

$$\text{Intensity} = i = \frac{\text{total energy radiated}}{\text{area of sun}} = \frac{E}{4\pi r^2}$$

where E is the total energy radiated per second. Since all this energy will strike the larger sphere of radius R, the intensity at the greater distance will be

$$\text{Intensity} = I = \frac{E}{4\pi R^2}$$

Now let's divide the two equations.

$$\frac{I}{i} = \frac{r^2}{R^2}$$

This equation, known as the inverse-square law, tells us that the intensity at any surface varies inversely as the square of the distance from the radiating body.

The intensity of the sun's radiation at the surface of the sun is then

$$i = I\frac{R^2}{r^2} = 1.94\left(\frac{93,000,000}{433,000}\right)^2$$

$i = 89,000$ calories per square centimeter per minute, or
$i = 1,480$ calories per square centimeter per second
This figure of solar intensity can be used with the Stefan-Boltzmann law to give the effective temperature of the sun's surface.

$$T^4 = \frac{1,480}{1.36 \times 10^{-12}}$$
$$T = 5700° \text{ K}$$

The temperature is very close to the accepted value of 5750° K and agrees well with the temperature obtained with Wien's displacement law.

The kind of light radiated by the sun is a result of the temperature of its surface—5750° K. If its surface were hotter, the maximum intensity would occur at the shorter wavelengths corresponding to blue, or perhaps violet. If the surface were cooler, the peak intensity would occur at the longer wavelengths at the

229

red end of the spectrum. And of incidental interest is the fact that human eyes have their greatest sensitivity near the wavelength of peak solar radiation, a fact of no great surprise since sight evolved in sunlight.

How are distances to the closer stars measured?

The distance to a relatively close heavenly body, such as the moon or one of the planets, is determined by measuring the direction of the body from different places on the earth. In general, the term *parallax* is used to signify the difference in direction of the object when it is viewed from different places. The more distant the object, the greater must be the separation between the points of observation.

To illustrate what parallax really means, hold up one finger about a foot in front of one eye and close the other. Now move your head from side to side slightly and you will see your finger "move" back and forth, or *oscillate*, with respect to the wall on

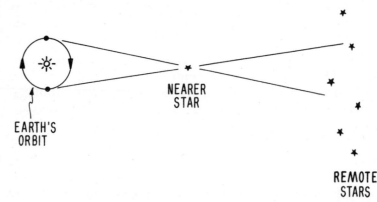

FIGURE 44. Parallax is the difference in direction of a star when viewed from different places.

the other side of the room. When astronomers make precise measurements of the locations of near-by stars, they find that they oscillate back and forth during the course of a year with respect to the more distant stars in the celestial sphere. As you can see from Figure 44, a nearer star will seem to be located in different places in the heavens depending upon the location of the earth in its orbit. The extent of this displacement is determined by taking photographs at 6-month intervals and noting the position of the nearby star with reference to others in the photograph. Part of this displacement, however, may be a result of the actual, or *proper,* motion of the star itself. For this reason, astronomers take many photographs at half-year intervals in order to disentangle the parallax oscillation from the proper motion of the star.

When used in a quantitative sense, the parallax of a star usually means the *heliocentric parallax*—or half the greatest parallax displacement that is observed for the star. Slight corrections are made to bring the earth to its average distance from the sun. The astronomical term *parsec* is the distance at which a star would have a *par*allax of one *sec*ond. One parsec equals 19.2 million million miles (19.2 $\times$ 10^{12} miles). The distance to a star in parsecs is equal to the reciprocal of its parallax in seconds. One parsec equals 3.26 light-years.

One of our nearest neighbors among the stars is the bright double star Alpha Centauri. Its parallax is 0.760 seconds, and its distance is $1/0.760 = 1.32$ parsecs. Its distance in light-years is $1.32 \times 3.26 = 4.3$ light-years. If we were to represent the sun by a dot on this page, Alpha Centauri would correspond to a pair of similar dots about 5 miles away.

This method is useful in measuring stars up to about 300 light-years distant from the earth. At greater distances, the parallax becomes too small to be measured reliably.

How was the size and shape of our galaxy determined?

Astronomers are able to measure the distance to the nearer stars—perhaps 200 or 300 light-years away—by the so-called *parallax effect* discussed in the previous question. But this method is hopelessly ineffective for stars as far distant as the end of our own galaxy, the Milky Way. Luckily, nature has distributed throughout the heavens a large number of extremely bright stars known as Cepheid (pronounced se-fee-id) variables, that make it

231

possible to measure the distance to extremely remote stars. Cepheids are easily identified because the intensity of their light varies periodically in a characteristic and predictable way. Such stars undergo a rather rapid increase in brightness followed by a somewhat slower decrease in brightness, this process being repeated over and over again at a perfectly definite rate. The best-known example of the Cepheid variable is the star Delta Cephei which has given the name to the class. This particular star somewhat more than doubles its light output in about 24 hours and then decreases in intensity over the next 4 days to its original brightness. Then it quickly brightens again to be followed by the same slow fade. The entire rhythmical cycle, or *period*, takes just about $5\frac{1}{3}$ days and the period is unchanging from each cycle to the next.

But other Cepheids have different light changes and different periods. Some brighten up as much as six times, and some have periods ranging from less than 1 day to as long as 2 weeks. It is in these variations that the Cepheids provide their distance-measuring properties: the longer the period, the brighter the star. And what's more, there's a perfectly definite relation between a Cepheid's period and its absolute brightness. If the absolute (or intrinsic) brightnesses of a number of Cepheids are plotted against their periods, they all lie on a simple curve (Figure 45). Observing such stars over a length of time provides their period, and once this is known the absolute brightness can be read off the curve. This is then compared with how bright the star *seems* to be. Naturally, the more distant the star, the less bright will it appear to be. The difference between a star's absolute brightness and its apparent brightness is a measure of its distance.

The discovery of this property of the Cepheid variables was first made by studies of the Magellanic Clouds, two irregular clouds each containing hundreds of thousands of stars. Hundreds of stars in the smaller cloud were found to be Cepheids. In addition, this cloud lies just outside our own galaxy and is sufficiently compact, astronomically speaking, for all its stars to be considered an equal distance from the earth. Even though they are very far

apart, their separation is small compared to their distance from
the solar system. This makes it possible to neglect the distance
involved, since it affects the brightness of all stars equally. If one
star in the cloud *looks* twice as bright as another, then it *really is*

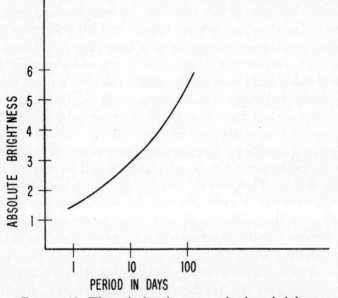

FIGURE 45. The relation between absolute brightness
and the rate of light variation of Cepheid variable stars.

twice as bright. In this way, many Cepheids were measured and
the principle of the period-luminosity law developed. Then a few
of the nearer Cepheids were measured and their distances determined by other means. This gave the absolute brightnesses of
these stars of known periods. The curve was then tied down completely and could be used on other Cepheids throughout the
galaxy.

A knowledge of the direction and distance of these stars gave
us the general size and shape of our galaxy. We now know it to
be a flattened disk some hundred thousand light-years in diame-

ter. The sun is close to its central plane about thirty thousand light-years from the center.

Which is the most important of all the numbers?

Reckoning was undoubtedly one of the earliest arts practiced by man. During 5,000 years prior to the birth of Christ, man developed a heritage of literature, art, philosophy, and religion. Many civilizations rose to great heights and fell before new conquerors. But over this entire span of time man could only develop such a crude and complicated system of numeration that experts alone could manage the simple calculations that we expect of children today. Their adding machine, the abacus, was not a convenience that was resorted to for the sake of speed and simplicity; it was an absolute necessity. During this entire period of time, not one single worthwhile improvement to the instrument or to the principles of calculation found its way into the minds of men. Then one day during the early centuries of the Christian era an event of world-shaking proportions took place—an unknown Hindu discovered zero! Many mathematicians agree that without zero the civilization that we know would never have gotten off the ground.

The clumsy number systems of the Greeks, Romans, and other early peoples used letters of the alphabet to represent each column of the counting frame, or abacus. The Romans, whose system is still familiar to us, used the letters I, X, C, and M for the columns of units, tens, hundreds, and thousands. If a letter were repeated 3 times it meant that there were 3 beads in the appropriate column. Thus, XXXII meant 3 beads in the tens column and 2 in the units column. In such a system, the position of the letter in the composite number has no significance. Translated into the arabic numbers that we use today, XXXII = 32. Notice that we can use the same digits (3 and 2) to mean 32, 302, 2003, and many others merely by adding zeros and *changing the location of the digits*. The Romans might have written 32 as (III)(II) and 23 as (II)(III), but it seems that nobody thought of doing it. If someone had, it would have been a long step in the right direction.

The beautifully simple idea of expressing all numbers by means of 10 symbols ultimately came to us from India. It probably arose from the necessity of recording the results of computations made on the counting frame. Suppose that an unsophisticated Indian had just calculated the number of coins that his master had received during the day. And suppose that there were 3 beads in the hundreds column, 0 in the ten column, and 2 in the units column. It would be simplicity itself to record the day's receipts as $\equiv 0 =$. The first digit $\equiv$ would mean 3 beads in the hundreds column, the 0 would mean that the tens column was empty, and the $=$ would stand for 2 beads in the units column. Naturally, it was only a matter of time for the $\equiv$ to become $\gtreqless$ and for the $=$ to become Z, and the number system as we know it began to take shape. Helping to support this theory is the fact that the Hindu word for 0 is *sunya*, meaning *empty*.

In this way, a number system seems to have evolved which gave to each symbol a value which depends upon its position in the number as well as its absolute value. This is one of the most profound mathematical discoveries ever made, in spite of its apparent simplicity. It changed the most difficult calculations into simple ones and laid the foundation for the invention of arithmetic—the invention that released man's mind once and for all from the shackles of the abacus. And it gave us a number system that would stretch out to infinity in either direction.

What is an abacus?

Man developed a need for written numerals long before rapid and simple calculation seemed important. Consequently, his number script did not lend itself to the performance of simple arithmetical operations. As the numbers he was forced to use grew larger and larger, it became necessary to rely on a sort of ancient adding machine called an *abacus*, or *counting frame*. As we learn more and more about numbers and their manipulation, we realize how much of a burden this really was. Man's ability to rea-

235

son mathematically was eventually limited by the rigidity and lack of scope of his material equipment. Clever as it was, the abacus was a mental crutch that hampered man's mathematical growth.

To discover how it might have come into existence, let's consider the problem of the ancient shepherd. As the flock grew larger, the head of the family probably sent a number of his sheep out to pasture under the care of one of his sons. But how can the older shepherd be sure that the same number of sheep that leave in the morning are returned at night? He probably solved this problem by sending the sheep out one by one, adding a pebble to a pile as each sheep went through the gate. That evening when the flock returned, he could reverse the process, removing a pebble from the pile as each sheep filed back into the enclosed area. If any pebbles were left over, he could take appropriate disciplinary action against the offending young shepherd. If there weren't enough pebbles in the pile to take care of all the sheep, he knew he must sleep close to his war club that night in order to deter any angry neighbors that might come calling. In any event, a method of tallying had been developed. Such a practice would lead inevitably to the counting frame. At first it might have been a row of parallel grooves in a flat surface such as a piece of wood or clay. Later it might have become a group of upright sticks on which could be placed a number of pierced stones or beads. Finally, the row of sticks became closed at both ends with a fixed number of beads (usually 10) in each row.

The counting frame seems to have been developed at very early times by many civilizations around the world. The Mexican and South American Indians were using it when Columbus "discovered" America. The Chinese and Egyptians had developed a form of the instrument thousands of years before the Christian era. Until that time, it was the only means by which calculation could be carried out.

The number system that we use today consists of symbols which enable us to perform the various arithmetical operations. To us the symbol is the number. This seems so simple that we under-stand it without even bothering to realize that we understand it. But the concept of numbers and symbols is comparatively new. The ancient number scripts such as were used by the Greeks and Romans were merely labels used to record the results of work done with an abacus (Figure 46).

In one form of the abacus, there are four parallel rows of beads. Each row consists of 10 beads. The first row is the units column, the second is the tens column, the third is the hundreds column, and the fourth is the thousands column. Suppose we want to add 58 and 63. We start by pushing all the beads to the right. Then

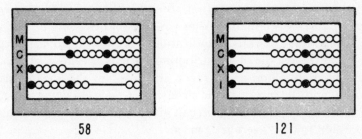

58 **121**

FIGURE 46. An abacus set for 58 and another set for 121.

we put the number 58 into the abacus. Eight beads are moved over to the left of the units column, and 5 beads are moved to the left of the tens column. Now we must add in the other num-ber, 63. This is done by attempting to move 3 beads to the left of the units column. When we have moved 2 of the three over, we find that the units column is out of beads, so we move 1 bead over to the left of the tens column and return the 10 beads to the right of the units column. We can now move our third bead to the left of the units column. Now we must add in the 6 of our second number. When we try to do this, we find that there are only 4 beads on the right side of the tens column. This means that we must move the 4 beads over to the left, move 1 bead to the left of the hundreds column, and return the 10 beads to the right of the tens column. Now we can move the last 2 to the left

237

of the tens column. A Roman mathematician would give the answer as CXXI. Complicated? Well, just reflect that the only alternative was to write the sum as 121 I's in the Roman system. This would have had as much significance to the Romans as the national debt of the United States has to us.

When is an average not an average?

There's an old adage to the effect that "figures don't lie, but liars figure!" The art of lying with statistics has become an efficient and honored ocupation in certain disreputable circles, so the rest of us had better learn some of the techniques in self-defense.

A statistical tool that is often abused is the so-called *measure of central tendency*—a quantity that you and I would call the average. The average gives us an indication of the way in which scores pile up at the middle of a distribution. Strictly speaking, at least three different kinds of averages are possible: the mean, the median, and the mode. If one's conscience isn't too demanding, use of the appropriate average can give a completely misleading impression to the "average" person.

The *mean* is merely another term for the arithmetic average. It is computed by taking the sum of a number of scores and dividing by the number of cases. If we have five scores, 10, 20, 30, 40, and 50, the mean is

$$\frac{10 + 20 + 30 + 40 + 50}{5} = 30$$

The *median*, on the other hand, is defined at that point in the distribution which has 50 per cent of the *cases* on either side of it. In the example given above, the median is also 30, since there are two higher and two lower scores in the distribution.

The *mode* is defined as that score which occurs most frequently. Suppose we were to ask 100 persons which of 10 brands of cigarettes they prefer. The results might look as follows:

238

	Number
Brand *A*	6
Brand *B*	7
Brand *C*	10
Brand *D*	13
Brand *E*	14
Brand *F*	15
Brand *G*	13
Brand *H*	10
Brand *I*	7
Brand *J*	5

The mode, then, is 15, since brand *F* occurs most frequently.

To illustrate some of the diabolical techniques that have been developed in the use of averages, suppose that one wants to influence the sale of brand *B* cigarettes. It's well known that brand *B* is far from being the most popular one, so a bit of chicanery is in order. How can one prove that most persons smoke brand *B* even though it's a fact that brand *F* will usually win in a fair fight? Here's how it's done. Merely select a group of about 5 persons at random and ask their preference. You will undoubtedly lose on the first attempt, so discard those irrelevant data and try again. With fortitude, you can question hundreds of groups of 5 persons in a day and you can be *sure* that one of these will turn out as follows:

Brand *B*	2
Brand *F*	1
Brand *G*	1
Brand *I*	1

You can then proclaim to the world, that "in a recent test brand *B* was preferred 2 to 1!" Nothing need be said concerning the hundreds of "irrelevant" tests that gave negative results.

Similarly, the treasurer of an organization may want to boost contributions through a judicious use of statistics. Suppose that last year's contributions were as follows:

239

$100.00
3.00
3.00
2.50
2.00
2.00
1.75
1.50
1.25
1.00
———————
$118.00

In his fund-raising speech he can state that the "average" contribution last year was $11.80, adding, "How about yours?" He means, of course, that the *mean* was $11.80—and he knows full well that most of it came from their one wealthy member! The more accurate picture is presented by the *median* which, in this instance, is $2.00.

All this merely serves to emphasize that statistics, when handled in a careless (or careful) manner, can lead to incredibly erroneous conclusions. When statistics are used in scientific expositions, most scientists prefer to see the "raw" data for themselves, so they may form their own conclusions. It's not that they mistrust their colleagues, it just is that the interpretation of statistics has been much abused for much too long.

What are degree days?

Have you ever wondered how fuel-delivery companies know when to deliver fuel without being called by the consumer? Heating engineers have solved the problem through an ingenious principle known as the *heating degree day*.

There is a close relationship between the amount of fuel used in a heating system and the number of degrees that the mean daily temperature falls below 65° F. The term *heating degree day* is merely a reduction of one degree of the mean daily temperature below 65° F. For example, if the mean temperature on a certain day was 40° F, that day would have accumulated 65 —

40 = 25 degree days. If the following day's mean temperature was 15° F, 50 degree days would have been accumulated. The fuel used on the second day would have been twice as great as on the first. Detailed studies have shown a marked correlation between heating degree days and fuel requirements. Moreover, this correlation is accurate throughout a wide temperature range and may be used to estimate fuel requirements with great precision. The accumulation of degree days, therefore, is an indirect measure of fuel consumption and can be used to determine fuel-delivery dates. In much the same way, *cooling degree days* give an accurate idea of the fuel needed, and consequently the cost of running a cooling system. The same principle is involved except that the excess of the mean daily temperature above 75° F is the significant factor.

What is a sidereal day?

Like the sun, stars rise in the east and set in the west, each moving in a plane that is perpendicular to the earth's axis. It takes a star just one *sidereal day* to make one trip around the earth. So any one of the stars can serve as a timepiece. All we need do is imagine a line drawn from the stars in question to the North Star, Polaris. This line will be the hour hand of our star clock. A day in such a system is the interval between two successive transits of the star over the same meridian. (A star transits when it crosses the observer's meridian.) Since a star does so twice a day, astronomers distinguish between *upper transit* above the celestial pole and *lower transit* below the pole. It's noon when the time star is at upper transit and midnight when it's at lower transit.

The length of a sidereal day is a few minutes shorter than the length of the solar day. This is because of the rotation of the earth around the sun. When the earth is at one end of its orbit, a certain star will transit at (solar) midnight. Six months later, when the earth is at the other end of its orbit, the same star will transit at noon. Averaged over the year, this effect causes the solar day to be about 3 minutes 56 seconds longer than the sidereal day.

Is the earth's orbit really an ellipse?

We usually think of the earth's orbit as elliptical, although the earth really wobbles back and forth in an *approximately* elliptical path around the sun. Similarly, the moon doesn't really revolve around the center of the earth, as most of us believe, but about a point at some distance from the earth's center. All this is a result of a planetary game of tug of war played continuously by the earth and the moon.

This will become clear when we realize that the earth and the moon are more like a double planet than a primary and a satellite. Though not the largest satellite in the solar system, the moon is no pygmy in its class. Its diameter is 2,160 miles and its mass $\frac{1}{82}$ as great as the earth's. Of more importance, however, is the fact that the moon is more nearly comparable with its primary in size and weight than any other satellite. Imagine the earth and the moon to be connected by a stiff, unyielding rod. Such an earth-moon system would balance at a point called the *center of mass*. It is this point about which the earth and the moon mutually revolve and about which they play their game of tug of war.

When we speak of the earth's elliptical orbit around the sun, it is the path of the center of mass that we are really talking about. Because the earth is much the heavier of the two, the center of mass is only 2,900 miles from the center of the earth. So this point, while within the earth, is closer to its surface than to its center.

What is chance?

The expression "the laws of chance" would seem to present a paradox of considerable magnitude. In his *Calculus of Probabilities*, Bertrand begins, "How can we venture to speak of the laws of chance? Is not chance the antithesis of all law?" We think of chance as the opposite of certainty and, therefore, a subject about which we know absolutely nothing. And yet, insurance companies and race tracks continue to pay dividends and gambling houses continue to prosper.

Perhaps an example will help to give an insight into the nature of chance. Imagine a bottle of air with the imprisoned billions of molecules darting back and forth at great speed, colliding with each other and with the walls of the container. And suppose that you were asked to concentrate your attention on any one molecule and calculate its position at any future instant. You would soon conclude that the problem is hopeless since the mathematics involved would be beyond human ability. And yet, using the laws of chance, there's a great deal that can be said about that particular molecule. We know, for example, that on the average it will probably be located in the lower half of the bottle as often as in the upper. What is the characteristic of chance that makes this determination possible?

If you examine a few situations to which we usually apply the laws of chance you will find many small and complex causes producing a large effect. Consider the molecule once more. It must travel a very great distance compared to its diameter before it collides with another molecule. If the direction of its motion is caused to change ever so slightly, it will strike the molecule toward which it is moving at a different point, sending both off at angles radically different from those that would have occurred in the absence of the slight deviation.

When a pair of dice have left the player's hand, the number that will turn up has been determined—and no amount of exhortation will materially effect the outcome. The number depends on many factors, all contributing their share: the resiliency of the dice, their true shape, the material of the table, the force of the throw, and even the air movement produced by the hysterical outcries of the participants. In short, the number depends upon a multitude of small, complex causes contributing to a large effect—an effect that bears no observable relationship to the original positions of the dice.

We apply the laws of chance to these situations, not because they refuse to obey natural laws, but because we do not yet know how to apply such natural laws as may be involved. Science, in a sense, is an attempt to determine and apply natural laws to phenomena that were previously thought to be random. The great Poincaré wrote, "You ask me to predict the phenomena that will be produced. If I had the misfortune to know the laws of these phenomena, I could not succeed except by inextricable calculations, and I should have to give up the attempt to answer you; but since I am fortunate enough to be ignorant of them, I will give you an answer at once. And, what is more extraordinary still, my answer will be right."

All the many phenomena about which we are ignorant are of two kinds—those that are random and those that are not. The former are the result of innumerable small and complex causes which defy precise mathematical treatment but which obey the laws of chance. The latter comprise those phenomena which are not random, and about which we will remain in complete ignorance until the associated laws are determined.

What is the basis of our number system?

We have seen that the ancient Greeks became extremely proficient in geometry and in the art of deductive reasoning that produced it. But this very proficiency seems to have prevented the Greek mind from expanding beyond the circles, squares, and cubes of Euclid. Consequently, their number system was bogged down by predetermined notions concerning the physical world they saw around them. When it came to calculating, the Greeks were content to limit themselves to the use of the abacus. Indeed, as we have pointed out elsewhere, their number system made it impossible for them to do otherwise. Let's consider the number MMMCXXIII, which is 3123 in our arabic notation. The Roman "numerals" are merely written counterparts of the beads in the various columns of the abacus. Each of the four letters involved stood for a particular column of the counting frame. The number given above indicates that at a particular time there were 3 beads in the first or units column, 2 in the second, 1 in the third, and 3 in the fourth or thousands column. The modern version of the number gives the same information *without recourse to four different symbols*. Our arabic symbols have nothing to do with any one particular column of the abacus. They stand for the beads only. The particular column of the counting frame involved is always the same as the position of the symbol in the number. Empty columns are accounted for by the symbol 0. A little thought will show that our modern number system allows us to perform arithmetical operations on paper that previously had to be performed on a mechanical device. The number 3,123 can be added to any other number such as 4,325 merely by placing one below the other and adding up the "beads" in each column. If any one column happens to run over 9, we merely throw 1 bead over into the next higher column, a process called "carrying over" when I was a boy.

Further analysis of the new system reveals even more significant characteristics. Stifel, the first European to use the signs +, −, and $\sqrt{}$, wrote, "I might write a whole book concerning the marvellous things relating to numbers." One of the things he might have written about is the connection between the abacus and our number system. Take the number 1,234,567, for example. (I might have written any other number, but that one happened to strike my fancy.) If we have an abacus with seven columns, we could set that number into the instrument by placing 7 beads in the units column, 6 in the tens, and so on. Such an abacus would be capable of handling numbers up to the millions, because it has seven columns. But 1 million is really six 10's multiplied together. We may write 1 million as $10 \times 10 \times 10 \times 10 \times 10 \times 10$, or, if we prefer shorthand, 10^6. By the same reasoning, 100,000 is 10^5, 10,000 is 10^4, 1,000 is 10^3, 100 is 10^2, and 10 is 10^1. This tells us that each bead in the tens column stands for 10^1, each bead in the hundreds column stands for 10^2, and so on, for as many columns as we care to worry about. The number 1,234,567 really means $1 \times 10^6 + 2 \times 10^5 + 3 \times 10^4 + 4 \times 10^3 + 5 \times 10^2 + 6 \times 10^1 + 7 \times 10^0$. If the last, or units, column is worrying you, it will help to recall that any number raised to the zero power is 1—thus, $10^0 = 1$. If we care to put this in a simple rule, we may say that each bead in the nth column has a value of 10^{n-1}. Thus, each bead in the seventh column has a value of 10^6, or 1 million.

If you were to ask a Greek mathematician to state the physical significance of such numbers, he would probably have replied along these lines—10^1 stands for a line 10 units long; 10^2 stands for a square 10 units on a side; 10^3 stands for a cube 10 units on a side. Beyond this point, the exponent that we place in the upper right-hand corner of our number would have no real meaning in Euclid's geometry. In our new system, that exponent can be made as large as we please because we have a physical model of the number to fall back on. The number $10^{1,358}$ merely stands for an abacus having 1,359 columns, all of which are empty except for 1 bead in the very last column. Our number system, therefore, is based on the *position* of the symbol in the number as well as the absolute value of the symbol itself.

What are logarithms?

Suppose you were asked either to add two 17-digit numbers or multiply them; which would you elect to do? Probably the former, because addition is a much simpler arithmetical operation than multiplication—especially if the numbers are large. So a great reduction in effort is effected if we can reduce the multiplication of large numbers to a process of addition. *Logarithms* were invented to do just that. To illustrate the fundamental principle involved, let's write two series of numbers, one below the other.

1	2	3	4	5	6	7	8
3^1	3^2	3^3	3^4	3^5	3^6	3^7	3^8
3	9	27	81	243	729	2,187	6,561

The bottom series of numbers given above is equal, of course, to the second. The principle, which was known to Archimedes, is this: pick two numbers in the bottom series which you want to multiply, 27 and 243, for example. Add the corresponding numbers in the top series, $3 + 5 = 8$. Go to column 8 and read the answer as 6,561. In other words,

$$3^3 + 3^5 = 3^{3+5} = 3^8$$

or, in general terms,

$$a^m + a^n = a^{m+n}$$

where a is any base number, and m and n are any exponents. This principle tells us that we can convert multiplication into addition if we change the form of the numbers to be multiplied from M and N, to a^m and a^n, and if we happen to have a suitable table of numbers. If M and N are any two numbers to be multiplied we must set

$$M = a^m$$
$$N = a^n$$

For the sake of convenience, we can assume $a = 10$, since most logarithm tables are made up on this basis. Then

$$M = 10^m$$
$$N = 10^n$$

We can now define a *logarithm* in the following way:

$$\log_{10} M = m$$
$$\log_{10} N = n$$

The operator $\log_{10}$ should be treated as a mathematical noun which means *the logarithm, to the base* 10, *of.* Many textbooks omit the subscript 10 since the logarithm is understood to be of base 10 unless otherwise indicated.

Similarly, an *antilogarithm* is just the reverse of a log.

$$\text{antilog}_{10} \, n = N$$

In order to multiply M by N using logs, we must perform the operations indicated in the following equation:

$$M \times N = \text{antilog} \, (\log M + \log N)$$

This equation asks us to look up $\log M$ and $\log N$ in a table of logarithms, add these two numbers together, and find the antilogarithm of the sum by reversing the way we use the tables.

In practice, there are some complications involved in the use of logarithm tables which ought to be discussed before we leave the subject. First of all, a logarithm usually consists of a number having a whole number portion followed by a decimal fraction.

247

The logarithm of 280, for example, is 2.4472. The logarithmic tables, however, will give you only .4472. You must get the 2 on your own initiative. The 2 in this case is called the *characteristic* of the logarithm, and the .4472 is called the *mantissa*. In order to determine the characteristic, merely count the number of digits to the left of the decimal point in the number, and subtract 1 from that amount. The number 280 has three such digits, so the characteristic of its logarithm is 2.

To illustrate the use of logarithms, let's multiply 280 × 43.

$$\log 280 = 2.4472$$
$$\log 43 = \underline{1.6335}$$
$$\log 280 + \log 43 = 4.0807$$

$$\text{antilog } 4.0807 = 12{,}040 = 280 \times 43$$

The first table of logarithms was compiled by Briggs in the 1631 edition of *Logarithmall Arithmetike*. Briggs couldn't keep himself from pressing their value "... by them all troublesome multiplications and divisions are avoided and performed only by addition instead of multiplication and subtraction instead of division." The world has come a long way since 1631, however, and today even the use of logarithms is being superseded by mechanical and electronic machines. These machines can not only multiply and divide, but they can solve difficult equations and guide a missile around the moon. Technological progress, it seems, will eventually give to logarithms a place beside the abacus in mathematical oblivion.

What is the curvature of the earth?

In Figure 47, the vertical line *CD* represents the diameter of the earth. The chord *AB* crosses the diameter at right angles and represents the distance, 2*a*, between the points *A* and *B*. To find the curvature of the earth, we are interested in computing the "earth bulge," *b*, corresponding to a distance of 1 mile from *A* to *B*.

A well-known theorem of geometry tells us that *a* is the geometric mean between *b* and *c*.

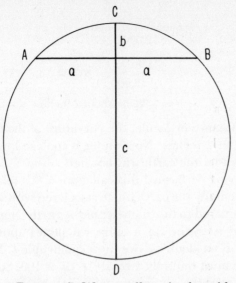

FIGURE 47. Water collects in the middle of long straight tunnels.

$$\frac{b}{a} = \frac{a}{c}$$

$$a^2 = bc$$

Since the distance $AB = 2a$,

$$a = \frac{AB}{2} = \frac{1}{2}\,\text{mile}$$

Then

$$a^2 = (\tfrac{1}{2})^2 = bc$$

In the present example, b will be extremely small compared to c so we can allow c to equal 8,000 miles, the approximate diameter of the earth.

$$b = \frac{1}{c}\,\left(\frac{1}{2}\right)^2$$

$$b = \frac{1}{8,000}\,\left(\frac{1}{4}\right)$$

249

$$b = \frac{1}{32{,}000} \text{ mile}$$

$$b = \frac{5{,}280 \times 12}{32{,}000} \text{ inches}$$

$$b = 1.98, \text{ or about 2 inches}$$

So in a distance of 1 mile, the curvature of the earth amounts only to about 2 inches! No wonder it took so long to convince the skeptics that the earth was not flat!

Once having performed this calculation, it's relatively easy to compute mentally the earth bulge associated with any chord that is small compared to the diameter of the earth. Notice that in the equation, $a^2 = bc$, the earth bulge b depends upon the *square* of the term a. If we double a, we must quadruple b. If we multiply a by 10, we must multiply b by $10 \times 10 = 100$. So to determine the bulge above the center of a 10-mile tunnel, merely multiply 2 inches by the square of 10.

$$2 \times 10 \times 10 = 200 \text{ inches} = 17 \text{ feet (approximately)}$$

The center of such a long, straight tunnel is 17 feet below the surface. Looking at it from another point of view, the center of the tunnel is 17 feet closer to the center of the earth than are its ends. Any water that collects in the tunnel will flow toward the center and remain there. A railroad train would travel downhill for the first half of its journey and uphill for the second half! This will not seem too strange when we consider that down and up are not directions like east and west which can go on forever. Down and up have meaning only with respect to a definite point —the center of the earth. In that regard, they are more like the directions north and south which are finite in nature. If we want our tunnel to be free of water, we will have to get a pump.

Why are multistage rockets used in space vehicles?

The most important formula in all rocketry determines the velocity that a rocket can attain after having burned all its fuel:*

* For a derivation see, for example: Arthur C. Clarke, *Interplanetary Flight* (New York: Harper & Brothers), p. 22.

$$V = C \log_e R \qquad (1)$$

or
$$V = 2.30C \log_{10} R$$

where V is the rocket's velocity at burnout, C is the speed of the exhaust or jet gases, and R is the "mass ratio" which is defined as:

$$\text{Mass ratio} = \frac{\text{initial mass of rocket}}{\text{final mass of rocket after burning fuel}}$$

In order to produce the fastest possible single stage rocket, then, we would want the greatest possible speed for the exhaust gases and highest mass ratio that the technology will allow. Let's investigate these requirements more closely.

When a chemical fuel burns, its chemical energy—the energy that's evolved during combustion—is converted into the kinetic energy that drives the gases out of the engine. In most rocket engines that we know about, this energy conversion takes place at considerably less than 100 per cent efficiency, a figure of perhaps 60 per cent being considered good. From such factors as these, engineers are able to select the fuel that should give the greatest over-all performance in terms of C, the jet velocity. One of the better fuels consists of alcohol and liquid oxygen, the fuel used in the V-2 and Viking rockets. Data from the flights of the Viking rocket indicate that an exhaust velocity of about 1.2 miles per second was attained. We also know that the Viking's weight breakdown is as follows:

	Per cent
Fuel	80
Structure	15
Payload	5
	100

Since the propellant constitutes four-fifths of the total mass* of the rocket, the mass ratio $= \frac{100}{20} = 5$. The theoretical velocity for the Viking is then:

* The "weight" of a rocket varies with its height from the earth and so has little meaning. The mass, or amount of matter present, is independent of the rocket's position.

$$V = 2.30(1.2)\log_{10} 5$$
$$= 2.30(1.2)(0.7)$$
$$= 2.0 \text{ miles per second}$$

Since the velocity needed to place a satellite in orbit around the earth is about 5.0 miles per second, the Viking's 2 miles per second is not to be taken lightly. Unfortunately, however, not all this theoretical velocity is realized in practice. Two factors contribute adversely to the final velocity attained by any rocket: air resistance and gravity.

Most of the loss of velocity owing to air resistance takes place in the lower atmosphere where the air is comparatively dense. As the rocket gains altitude, the air drag steadily diminishes. At an altitude of 20 miles, 99 per cent of the atmosphere is below the rocket and it travels in a relatively good vacuum. It is estimated that the Viking's loss of velocity resulting from atmospheric drag amounts to 500 miles per hour, (0.14 mile per second) over the course of its upward journey. While this is a regrettable waste, the loss resulting from the action of gravity is even more important.

If the rocket engine were the only force acting on a space vehicle, its final velocity at burnout would follow equation (1) very closely. But in addition to the rocket engine, we must also consider the impeding action of gravity. All during its powered ascent, the rocket must fight against the attraction of the earth. This produces a reduction in its velocity* at burnout equal to $g \times t$ where g is the acceleration due to gravity, 32.2 feet per second per second, and t is the time of powered ascent. Since the Viking's engine is operative for 100 seconds, the velocity loss is $32.2 \times 100 - 3,220$ feet per second, or 0.61 mile per second. Taking gravity and wind resistance into effect, the velocity of the Viking at burnout of its engine is $2.0 - 0.14 - 0.61 = 1.25$ miles per second. This is considerably less than 5 miles per second needed for an earth satellite, or the 7 miles per second required of a vehicle that is to reach the moon or orbit around the sun.

---

* Assuming vertical motion during powered ascent.

Increasing the size of a rocket such as the Viking will have little effect on the final velocity attained because the weight breakdown and hence the mass ratio will be substantially unchanged. Increasing the velocity of the exhaust gases may be possible to some degree through the use of so-called "exotic" fuels, but a four- or fivefold increase seems impossible to the experts. The only solution to the problem of attaining greater speed seems to lie in the use of multistage vehicles.

Referring again to the weight breakdown of the Viking, why not let the pay load, which constitutes 5 per cent of the total, consist of another rocket? Our space vehicle would then consist of a large first-stage rocket with a smaller second-stage rocket mounted on top of it. When all the fuel of the first-stage rocket has been expended, the second rocket will be moving at 1.25 miles per second and can carry on from there, adding another 1.25 miles per second to the speed it already possesses. Actually, it will do much better than that. In the rarified upper atmosphere the second-stage rocket engine will be considerably more efficient because the exhaust gases will have a vacuum in which to expand. Neither will it lose any velocity to air resistance. Because of these factors the two-stage combination would probably reach a velocity close to 2.9 miles per second. The price paid for this increased velocity is a great reduction in actual pay load. The true pay load of a two-stage rocket designed along the lines of the Viking would be $\frac{1}{20} \times \frac{1}{20} = \frac{1}{400}$, or about one-quarter of 1 per cent of mass of the vehicle! In a three-stage rocket of this kind the pay load would be a minuscule one eight-thousandth of the total! In modern multistage systems such as the Vanguard vehicle a figure of 1,000:1 is probably closer to the truth, but even so, each pound of useful pay load represents half a ton of initial weight.

The only alternative to the multistage technique is quite impracticable at the present time. To put a satellite in orbit with conventional fuels, a single-stage rocket would require a mass ratio of about 50:1, which would seem to be next to impossible. Such a ratio would require that 98 per cent of the vehicle's total mass be propellant. Only the remaining 2 per cent would be available for pay load, structure, rocket engines, and the like.

253

Before leaving the subject, a few useful and easy-to-remember relationships can be deduced from equation (1) when it is arranged in a slightly different form.

$$V = C \log_e R$$

$$\frac{V}{C} = \log_e R$$

$$R = e^{\frac{v}{c}}$$

The number $e = 2.718$ is the base of natural logarithms. If $V = C$ in the final equation, $R = e^1 = e = 2.718$. In other words, if the mass ratio is equal to 2.718, the final velocity of the rocket, V, equals the velocity of the exhaust jet, C. If $V = 2C$, $R = e^2$. For a rocket to travel *twice* as fast as its exhaust gases the mass ratio must equal $(2.718)^2 = 7.39$. To triple the speed of the exhaust jet a rocket would need a mass ratio equal to $(2.718)^3 = 20.1$. It would appear, therefore, that the practical limit for the speed of a rocket is somewhere between two and three times as fast as its exhaust.

Does every day have 24 hours?

This is a deceptively simple question. When we speak of a day, we usually mean a *solar day*—the time it takes the earth to rotate from noon of one day to noon of the next. Since there are 24 hours involved in such an operation, we might hastily jump to the conclusion that the answer to the question is "yes"—but we would be wrong. To illustrate, suppose we were to perform the following experiment: Let's wait until the shadow of the sun points directly north and set an accurate clock to 12 noon. Then after approximately 24 hours have passed, let's wait until the sun's shadow again points due north and read the time on the clock. If we perform the experiment with accuracy, we will find that the sun will run as much as 10 or 15 minutes fast or slow, depending upon the season of the year. During November it will be up to 16 minutes fast, while in February it will run about 12 minutes slow.

The sun that we use for keeping ordinary time is the *mean sun*, or *average sun*. *Mean solar time* is time by the mean sun. We wouldn't have to resort to averages in time determination if the rotation of the earth were precisely uniform; its day would then be of constant length. But it turns out that the earth is continually slowing down and speeding up as it moves around the sun. This results from the second of Kepler's laws concerning the motions of the planets around the sun. This law states that *the line joining a planet to the sun sweeps over equal areas in equal lengths of time*. The nearer a planet comes to the sun, the faster it moves because it must travel a greater distance to sweep out an equal area in a given length of time. Since the earth is closest to the sun during winter (in the Northern Hemisphere) it must be traveling fastest. During summer, the longer line joining the earth and sun need not go around as far as the shorter line in winter to sweep out the same area.

But perhaps you're wondering what all this has to do with the length of a day. After all, when we talk about the length of a day we are concerned with the rotation of the earth *about its axis, not around the sun*. As you will see, the two are quite definitely related. Imagine that you're located high above the sun looking down at the earth as it rotates around the sun. Now pick out a spot on the earth which is directly in line with the sun. The solar time at this point is precisely 12 noon. But the earth is spinning like a top as it moves slowly in its orbit. During the time it takes to spin around once, it will have moved *about* 1° in its orbit. So the point that we selected for 12 noon "yesterday" has not quite come around in line with the sun. It will take a few minutes longer to do so. But the earth revolves farther in a day in winter and, therefore, must rotate farther to again bring the sun directly overhead. For this reason, days by the sundial are longer in winter than in summer.

In addition to its poor qualities as a time piece, it appears that the solar clock is actually running down. Astronomers have calculated the time when certain eclipses should have occurred in history, and have found that the recorded times are not in agreement with their figures. Such calculations have caused them to conclude that the length of the day is increasing at the rate of 0.0016 second each century. This means that the solar clock has

255

run slow about 3½ hours during the past 2,000 years. Scientists believe that the ocean tides are producing the braking action which is slowing down the earth.

Is there an easy way to estimate angles?

Oddly enough, most of us have a fairly accurate angle measuring device that's "built in." This natural protractor consists of one eye, the length of an arm, and the width of a hand. Here's how it works. Close one eye and extend your arm out in a horizontal direction. Now tilt your hand so that the palm is away from you. At the second joint, your hand is now measuring an angle of very close to 8°! It will vary somewhat from person to person depending upon the length of arm and width of hand, but it usually turns out to be very nearly 8°.

You can "calibrate" yourself in a few minutes, if you like, and never be in doubt as to the "accuracy" of your own protractor. Merely stand near the corner of a room and measure the number of hand widths it takes to make up a right angle. If you measure 11 hand widths at the second joint (88°), you have an 8° hand. Other results will, of course, require slight modifications to this figure.

Artillery officers use a slight variation of this angle-measuring technique to good advantage. A *mil* is an angle which measures one foot at a distance of 1,000 feet, or one yard at 1,000 yards, or one meter at 1,000 meters. This is a convenient angle because your hand can be calibrated quite accurately in mils. The width of the index finger measures 40 mils; the second finger, also 40 mils; the third finger, 35 mils; and the little finger, 30 mils. Your four fingers taken together measure 145 mils. You can use this information in many ways if you happen to be curious about hard-to-measure distances. Suppose that you want to know the height of a tree, or a chimney, or a building, or a smokestack. All that you have to do is walk 100 yards away from the object, turn your outstretched hand sideways, and measure from the ground to the top of the object. If it takes 80 mils (the first two fingers) the height is 8 yards. If your measurement had been made at 1,000 yards, the height would have been 1 yard per mil of height, or 80 yards. Since you were only 100 yards from the object, its height is $\frac{1}{10} \times 80 = 8$.

Why do we call irrational numbers "irrational"?

In ancient times men spoke of distances, as we do, in terms of units of length. The idea of a common measure was the basic tool by which they reasoned through problems in what we now call arithmetic. If a problem involved a rectangle, they would draw a diagram having a height of so many units and a length of a different number of units. In this way, areas and perimeters could be determined quite nicely. As time went on, mathematicians became more and more proficient in the use of their spatial arithmetic until—to their horror, I imagine—they discovered that there are many distances that do not possess a common measure! Take a right triangle 2 feet in length and 1 foot in height, for example. As we will see, the third leg of this triangle cannot be expressed *evenly* in the same unit of length as the other two sides. To understand this paradox, let's consider the 3-4-5 triangle. This, you will recall, is a right triangle having sides in the ratio 3:4:5. As you can see in Figure 48, if you erect a square on the base and another on the height, the area of these two squares (16 + 9) equals the area of a square erected on the hypotenuse (25). The Egyptians had discovered this "magic triangle" by measurement and it was also known to the ancient Chinese. It was considered magical because all the numbers turned out pleasantly even. The square of side 3 is 9, the square of side 4 is 16, and the square of side 5 is 25, and 9 + 16 = 25. Now let's consider our triangle 1-2-?. The hypotenuse is the unknown element in this triangle. But the Greeks proved that the hypotenuse is the side of a square which is equal to the sum of the areas of the squares erected on the other two sides. That is, the square of the hypotenuse must equal $2^2 + 1^2 = 2 \times 2 + 1 \times 1 = 5$. The problem, then, is to find a number which when multiplied by itself equals 5. The Greeks spoke of this number as the *square root* of 5, and we write it $\sqrt{5}$.

Finding such a number would seem to be reasonably straight-

257

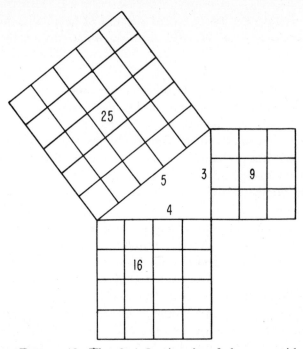

FIGURE 48. The 3–4–5 triangle of the pyramid builders.

forward assignment. Presumably, there must be a whole number plus a fraction of a unit which when multiplied by itself equals 5. But, unfortunately, there isn't. If you try $2\frac{1}{4}$ you will get $5\frac{1}{16}$. If you try $2\frac{1}{5}$ you will get $4\frac{21}{25}$. Testing $2\frac{2}{9}$ gives $4\frac{76}{81}$. You might go on trying any number of such combinations, but the square of any number-plus-fraction will always be another number-plus-fraction. It will never come out even. When men discovered this unpleasant characteristic of such numbers as the square root of 5 they referred to them—with pardonable irritation—as *irrational* numbers. The term has come down to us from early times as a reminder of the contempt in which such numbers were held. In any event, there is no common measure which can be divided into the sides and hypotenuse of such a triangle.

There is a bright side to this untidy situation, however. If irrational numbers did not exist, then we would have left only those numbers consisting of whole numbers and fractions. This would mean that numbers are not continuous. To illustrate, let's examine all the fractions from 0 to 1. You would expect there to be a limitless or infinite number of fractions between 0 and 1 that are so close together as to be continuous. But this is impossible. Two fractions must always be different, or else we would have no way of telling them apart. No matter how close together we pack our fractions, there must always be some tiny space between them. Luckily for our peace of mind, mathematicians have shown that irrational numbers fill the gaps between fractional numbers. If this were not true, it would be possible to draw a line 1 inch long and then mark off a portion of that line which could not be specified by any number. That would be enough to drive any mathematician to distraction!

How many regularly shaped tiles are possible?

Tilemaking is an ancient art and many interesting combinations of shapes and colors have been used through the ages. But if you think back, for a moment, you will recall that only three *regularly* shaped tiles are ever used—the triangle, the square, and the hexagon. There's a good reason for this lack of variety and, as usual, the reason was well known to the ancient Greeks. Of the infinitude of regular polygons (plane figures having equal sides and angles) only three can be made to fit together precisely with no gaps and no overlap: the triangle, the square, and the hexagon.

Naturally enough, the Greeks weren't content to have discovered this fact; they proceeded to prove it. The proof went something like this. Consider a regular polygon of K sides, a portion of which is drawn in part (a) of Figure 49. Let A be the vertex angle at each corner of the polygon. Then draw lines from each vertex to the center as in (b), producing K triangles, one for each side. Since the angles of each triangle add up to $180°$,

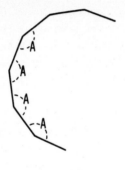 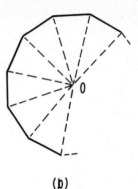

(a)   (b)

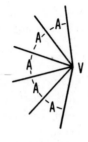

(c)

FIGURE 49. (a) A regular polygon having an indeterminate number of sides. (b) The same polygon with lines
drawn from each vertex to the center. (c) An indeterminate number of such polygons coming together at point V.

the polygon (consisting of K triangles) will contain $K \times 180°$.

Sum of all angles in polygon $= K \times 180°$

Of that amount, a certain number of degrees occur at the vertices
and the remainder occur at the center of the polygon O where all
the triangles meet. The latter amount to just $360°$ or, as the
Greeks would say, four right angles. The sum of all the vertex
angles, then, is $K \times 180° - 360°$ or $(K - 2)180°$.

260

$$\text{Sum of all vertex angles} = (K - 2)180°$$

Since there is just one vertex angle for each of the K sides,

$$\text{Each vertex angle} = A = \frac{(K - 2)180°}{K}$$

Now we must place a number of regular polygons together on a flat surface as shown in part (c) of the diagram. Since we have no idea of the number of polygons to use, let's allow N to stand for the requisite number. We now have N polygons placed together on a flat surface. A vertex of each polygon meets at point V in the figure and each contributes its vertex angle A at the point. Since there are to be no gaps or overlapping of the polygons, the number of vertices times the angle A must produce exactly $360°$. Since there is one vertex per polygon at point V,

$$N \times A = 360°$$

or

$$A = \frac{360°}{N}$$

The two expressions for A can now be placed equal to one another.

$$\frac{360°}{N} = \frac{(K - 2)180°}{K}$$

Dividing by $180°$

$$\frac{2}{N} = \frac{K - 2}{K}$$

Inverting,

$$\frac{N}{2} = \frac{K}{K - 2}$$

or

$$N = \frac{2K}{K - 2}$$

This expression can be made somewhat easier to interpret by adding and subtracting the same number, 4, from the numerator and simplifying.

261

$$N = \frac{2K - 4 + 4}{K - 2}$$

$$N = \frac{2(K - 2) + 4}{K - 2}$$

$$N = 2 + \frac{4}{K - 2}$$

The last equation tells the whole story. Since K and N must be whole numbers, several inferences can be drawn from the equation. First of all, K cannot be equal to 1 because N would then be a negative number. Similarly, the substitution $K = 2$ gives the indeterminate quantity $N = 2 + \frac{4}{0}$. Substitution of 3 and 4 for K gives possible solutions, but $K = 5$ runs into difficulty. If $K = 5$, then $N = 3\frac{1}{3}$; but N must be an integer, so the five-sided pentagon will not work. Substitution of $K = 6$ gives another possible solution, but all integers above 6 give impossible results. Seven, for example, gives $N = 2\frac{4}{5}$ which is impossible on two counts: first of all, N cannot be a fraction; secondly, N must be *equal to or greater than three*. If only two polygons are used, the vertex angle must be equal to 180° and the size of the polygon must be infinitely great. So all numbers for K greater than 6 give impossible results. The only regular polygons that can be used in tilemaking, then, are the equilateral triangle, the square, and the regular hexagon.

Having proved their point, the Greeks went on to show that the same reasoning can be applied to solid figures constructed with faces of regular polygons. The most common of the *regular polyhedra* is the familiar cube that is so widely used in the manufacture of sugar and dice. The proof is analogous to that given above.

Suppose we want to construct a regular polyhedron using a number of identical polygons of K sides. Whereas the tiles were required to fit together on a *flat* surface, the present situation requires that a number of adjacent polygons meet at a vertex to form a *convex* surface (Figure 50). Let that number of polygons be N, and let their angles be A, as before. It's intuitively apparent

262

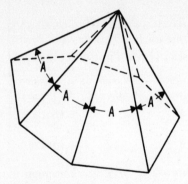

FIGURE 50. A number of polygons meeting at a point to form the vertex of a polyhedron.

that the sum of all angles meeting at a vertex of the polyhedron must be *less* than 360° (It's as though an angular segment had been removed from a piece of paper and the ends joined to form a cone.) Then, from the expression developed earlier, N must be less* than $2 + \dfrac{4}{K-2}$. Inequalities of this kind are usually represented as follows:

$$N < 2 + \frac{4}{K-2}$$

where the symbol $<$ is read "is less than." In the analysis of this expression, the solutions $K = 1$ and $K = 2$ are impossible for the reasons discussed in connection with plane figures. The solution $K = 6$ gives $N < 3$, and a little reflection will show that N cannot be less than three; if $N = 2$, only two polygons will be used at each corner of the polyhedron and the solid will become a plane figure with no depth. For the same reason, all solutions requiring K to be 7 or more are impossible.

It seems, then, that constructable polyhedra result when K is equal to 3, 4, or 5. We can now determine the corresponding numbers for N. If $K = 3$, then N must be *less than* 6. This means that N can be equal to either 5, 4, or 3 when three-sided regular polygons (equilateral triangles) are used. When $K = 4$, N must be less than 4, so only one solution is possible—$N = 3$. Similarly, when $K = 5$, N must be less than $3\frac{1}{3}$, so the solution is $N = 3$.

* Since $NA < 360°$

263

We find, from all this, that only five regular polyhedra are possible, three made of triangles, and one each of squares and pentagons. A synopsis of their characteristics is given in the table below.

Regular polyhedron	Number of vertices V	Number of faces F	Number of edges E	Shape of faces
Tetrahedron (pyramid)	4	4	6	triangle
Hexahedron (cube)	8	6	12	square
Octahedron	6	8	12	triangle
Icosahedron	12	20	30	triangle
Dodecahedron	20	12	30	pentagon

At first glance, the figures given in the three columns above do not seem to show any definite correlation. The relative number of vertices, faces, and edges that nature will tolerate in a regular polyhedron seems quite random. But René Descartes, the great French mathematician of the seventeenth century, noticed an unexpected relation between these quantities. An examination of the table will show that the sum of the number of vertices V and faces F always exceeds the number of edges E by 2. In algebraic shorthand,

$$V + F = E + 2$$

Whenever an equation of this sort pops up in mathematics, many minds set to work to find out why. Euler demonstrated a strict proof* of Descartes' relation showing that it is true, not only for regular polyhedra, but for *all* polyhedra—regular or not.

Why do artificial satellites travel so much faster than the moon?

It might seem odd at first glance to find that an artificial earth satellite completes its great-circle path around the world in about an hour and a half, while the moon takes almost a month to complete its circuit. Perhaps a simple analogy will help to clarify this point before we look into the matter quantitatively. Imagine a

* See, for example: Richard Courant and Herbert E. Robbins, *What Is Mathematics?* (New York: Oxford University Press, 1941).

stone tied to the end of a piece of string and whirling around in a circular path. If the stone is revolving fast enough, the tension in the string will balance the tendency of the stone to fly off and the system is in a state of balance. A little experimentation will indicate a definite relationship between the speed of the stone and the length of the string. As the string is made shorter, the stone must be moved more rapidly to keep it from falling to the ground. Similarly, a very long string will permit a relatively low speed to keep the stone in the air. In the case of a satellite, the force of gravity substitutes for the pull of the string. As the earth-to-satellite distance becomes greater, the satellite must move at a slower rate to keep the system in balance.

It turns out that there is a definite relationship between a satellite's speed and its distance from the center of the earth. For a circular orbit, the speed of a satellite is given by

$$V = R_e \sqrt{\frac{g}{R}}$$

where R_e is the radius of the earth, g is the acceleration due to gravity *at the earth's surface,* 32.2 feet per second per second and R is the distance from the satellite to the center of the earth. So as R becomes greater, the speed of the satellite decreases.

Another factor that must be considered, of course, is the length of the satellite's orbit, $2\pi R$. This quantity increases in direct proportion to the satellite's distance from the center of the earth. The higher the satellite, the longer its orbit and the slower its speed. Since the mean distance to the moon is about 60 earth radii, or 239,000 miles, it takes considerably longer for the moon to circle the earth than a satellite close to the earth's surface.

If a satellite is placed in orbit at an altitude of about 22,000 miles, its speed and orbital length are such that it will take just 24 hours to circle the earth. Such a satellite—placed in the equatorial plane—would remain constantly over the same point on the earth's surface. One of the first applications of such a "space platform" would be for communication purposes. Radio waves directed at the satellite could be rebroadcast to all points in a

265

hemisphere. Three such stations would make it possible to cover practically all the earth with a radio or television program.

Why is the day divided into 24 hours?

If we could erase the imprint of a lifetime of associations, very few of us would select the hour as a unit of time measurement. Why not a tenth of a day, or a twentieth, if we bring our toes into the picture? A clue to the answer is given by the ease with which the ancients were able to make an angle of 60°, and their demonstrated ability to bisect angles. Long before Galileo invented the pendulum clock, man had a need for the accurate measurement of the passage of time. This he accomplished by measuring the changing direction of the sun's shadow cast by a vertical pole or obelisk. This required the measurement of angles between various positions of the shadow. Of all the angles below 90°, one of the easiest to make is 60°. (An angle of an equilateral triangle.) Since ancient man found it easy to bisect angles, it was convenient to produce an angle of 15°, and that was the angle that he used between adjacent marks on his crude clock. But the earth rotates through 360° in one day, so that 15° would represent just one twenty-fourth of a complete revolution, the length of our hour.

Actually, the simple clock described above was not as accurate as it seems. When the sun rotates through 15°, the shadow of a *vertical* pole rotates through an angle that is not the same at all seasons. It depends upon the angle of the midday sun in the sky. So the length of an hour given by the shadow clock was not the same fraction of a day throughout the year. The Arabs finally solved this problem by the invention of the Moorish sundial.

What is probability?

A rigorous definition of probability might require several pages of careful groundwork, but for our purposes perhaps we can risk a direct statement. Suppose we are about to toss a coin. The result of the toss can be either a head or a tail. If we make a great number of tosses, and tabulate the ratio of heads to tails, then the probability that any one toss will be a head is the number ap-

266

proached by that ratio as the number of tosses tends toward infinity. That, you will agree, is a rather long-winded way of saying that the probability is 1/2 without actually saying it. This precaution is taken in order to have the definition include such items as loaded dice, two-headed coins, and the like. While the definition sounds mathematical, it's really the standard empirical definition of probability. We can arrive at a mathematical definition with the help of Pascal's triangle, an extremely versatile mathematical device.

If you toss an ideal coin once, you may get a head or you may get a tail. Let's represent this situation by the mathematical notation 1,1. If you toss the coin twice, there are 4 possibilities: you may get 2 heads; or 1 head followed by 1 tail; or 1 tail followed by 1 head; or 2 tails. Since the 2 middle possibilities are really the same, they may be grouped together and we have 1, 2, 1. If you toss a coin 3 times, you may get 3 heads; or 2 heads and 1 tail in any of 3 ways (HHT, HTH, THH); or 2 tails and 1 head in any of 3 ways (TTH, THT, HTT); or 3 tails. This situation can be represented by 1, 3, 3, 1. The 1 in this notation means that there is just 1 way of getting 3 heads (or tails) and the 3 means that there are 3 ways of getting the combination of 2 heads and 1 tail (or vice versa).

While it's by no means obvious, it turns out that these laws of probability can be found in the compact form of Pascal's triangle.

```
              1   1
            1   2   1
          1   3   3   1
        1   4   6   4   1
      1   5  10  10   5   1
    1   6  15  20  15   6   1
  1   7  21  35  35  21   7   1
1   8  28  56  70  56  28   8   1
1  9  36  84  126 126  84  36   9   1
```

Each number in the triangle is the sum of the two numbers above it. In the sixth row, for example, 15 = 5 + 10. While only nine rows are presented above, there is no limit to the triangle and it may be carried on *as long as we please.*

Now to demonstrate the use of this triangle. If we are going to make 3 tosses, we refer to the third row. Seven tosses would send us to the seventh row, 8 tosses to the eighth, and so on. Suppose we decide to make 5 tosses. The fifth row tells us that, starting from either end of the row, there are:

> 1 way to get 5 heads
> 5 ways to get 4 heads and 1 tail
> 10 ways to get 3 heads and 2 tails
> 10 ways to get 2 heads and 3 tails
> 5 ways to get 1 head and 4 tails
> 1 way to get 5 tails

There are no other possibilities. If we want to know the *probability* of getting 2 heads and 3 tails in 5 tosses, we reason as follows: There are 10 ways to get this combination; also, there are 32 *possible* ways to toss 5 coins, $1 + 5 + 10 + 10 + 5 + 1 = 32$. So the fraction $\frac{10}{32}$ describes the probability by giving the number of ways of succeeding (10) divided by the total number of ways the experiment (of tossing 5 coins) can come out. That is about as close as we can come to a mathematical definition of probability.

Can sound exist without a listener?

A mass of ice and snow crashes down the side of an antarctic mountain, far from the nearest ear of any kind. Was a sound produced by the avalanche? Debates of this kind can go on endlessly without hope of decision until a definition is arrived at for "sound."

Sound can be defined, in one sense, as an auditory sensation produced by an external disturbance upon the ear and its associated nerve endings. With this definition in mind, the avalanche could have produced no sound, for sound is then a purely *subjective response* that cannot exist without a listener.

On the other hand, sound can be defined as a vibratory disturbance in a transmitting medium which is capable (among other things) of producing an auditory sensation in an ear. This

268

definition clearly indicates that an avalanche certainly does produce a sound, since sound is considered a purely *objective phenomenon.*

Physiologists, otologists, acoustical engineers, and even psychologists are usually most interested in the hearing mechanism and the auditory sensation, while the physicist is most interested in the vibratory disturbance, whether or not a listener is involved. In order to avoid confusion, sound as defined by the former definition is usually referred to as *subjective sound* or *sound sensation,* in order to distinguish it from the *objective sound* or, more simply, the *sound* of the physicist.

The same problem arises in talking about the magnitude of a sound. The term *loudness* depends upon the ear as well as the physical properties of the sound, whereas the term *intensity* is used to describe the vibratory disturbance itself.

As with all types of wave motion, sound is energy in motion. The particles of air, or any other substance through which sound passes, act only as a carrier of this energy. Imagine a small sound source in air located far from physical objects that can reflect sound energy. Sound waves will move out from the source in all directions in the form of concentric spheres. The area of one of these spheres is $4\pi r^2$ where r is its radius. If the sound source is emitting energy at the rate of W watts, the intensity of the sound source is

$$I = \frac{W}{4\pi r^2} \quad \text{watts per square centimeter}$$

In ordinary conversation, a man emits about ten-millionths of a watt of acoustical power. If this power were to spread out uniformly in all directions, the intensity at a listener's ear 1 meter away would be

$$I = \frac{0.00001 \text{ watt}}{4\pi(100 \text{ centimeters})^2}$$

$$I = 0.00000000008$$

$$I = 8 \times 10^{-11} \text{ watt per square centimeter}$$

269

One microwatt is equal to 10^{-6} watt so the intensity given above is a miniscule 80 micromicrowatts per square centimeter. Small though it may be, this intensity is many hundreds of thousands of times greater than a sound barely audible to the normal human ear! According to acousticians, the threshold of audibility corresponds to about 10^{-16} watt per square centimeter.

In addition to intensity, most of the other quantities associated with sound are astonishingly small. As a sound wave moves along, it produces minute pressure variations in the intervening air. For an extremely weak sound, these pressure variations amount to less than a billionth of the normal atmospheric pressure, 14.7 pounds per square inch. Beating a sledge hammer on a nearby steel plate produces pressure variations only one-thousandth as great as atmospheric pressure. Similarly, the air particles carrying a very weak sound (1,000 cycle tone) move back and forth through a distance of only 7.4×10^{-9} centimeter—a small fraction of the diameter of the vibrating molecule itself. Because of this small distance, the resulting velocity of the molecule is a mere centimeter per day! A loud sound, however, will produce molecular velocities and displacements a million times as great.

In order to emphasize the unbelievably small quantities involved, M. Y. Colby[*] has calculated the length of time required to heat a cup of coffee by means of sound waves. If the cup is insulated perfectly against heat loss, and if the coffee is at room temperature, it will take about 2,000 years of conversation before the coffee would be hot enough to drink!

Why do the stars disappear in the daytime?

There is a well-accepted and long-standing concept, known as the Weber-Fechner Law, that seems to apply to all sense organs over rather wide limits. Applied to the eye, for example, the rule indicates that the eye can distinguish between two brightnesses if their *ratio* (not the difference between them) is greater than a definite and constant amount—one being about 5 per cent brighter than the other. Consider the disappearance of the stars in day-

[*] *Sound Waves and Acoustics* (New York: Henry Holt and Company, Inc., 1938), p. 206.

light, for example. The *difference* in brightness of the star with respect to its background is *always the same*—it always adds its bit of illumination to that of its surroundings—but the *ratio* in the daytime differs greatly from that at night. Expressed mathematically, the ratio is

$$\text{Ratio} = \frac{\text{background light} + \text{starlight}}{\text{background light}}$$

Even in the faint light of the full moon, hundreds of stars become invisible to the naked eye.

This dependence of human perception upon the *ratio* rather than the difference exists to a marked degree in the sense of hearing. We think of one sound being so many times as loud as another, whereas we would be hard put to say that the former is 43 micromicrowatts per square centimeter more intense than the latter. Like it or not, we live in a world of ratios.

Since ratios are so important, a valuable technique has been developed to handle them in a quantitative manner. The basic unit of measurement is the *bel* named after the American inventor, Alexander Graham Bell. The *bel* is defined as the logarithm (to the base 10) of the ratio of two amounts of power or energy. Let A and B designate two amounts of sound intensity that we want to compare. If n is the number of bels corresponding to the ratio $\frac{A}{B}$, then

$$n = \log \frac{A}{B} \text{ bels}$$

Through usage, the tenth of a bel, or *decibel,* has almost completely supplanted the *bel* in scientific work. Since 10 decibels = 1 bel,

$$n = 10 \log \frac{A}{B} \text{ decibels}$$

The abbreviation db is used universally for decibel, often in speech as well as in writing.

271

Because of the selection of base-10 logarithms, a few important ratios are easy to remember in decibel form. The table below shows how they are derived.

Ratio	Logarithm	10 × logarithm (approximate), db
1	0	0
2	0.30	3
4	0.60	6
10	1.0	10
100	2.0	20
1,000	3.0	30
10^6	6.0	60
10^{12}	12.0	120
$\frac{1}{2}$	−0.30	−3
$\frac{1}{4}$	−0.60	−6
$\frac{1}{10}$	−1.0	−10
$\frac{1}{100}$	−2.0	−20
$\frac{1}{1,000}$	−3.0	−30
$\frac{1}{10^6}$	−6.0	−60
$\frac{1}{10^{12}}$	−12.0	−120

Under the most favorable conditions, the average human ear can distinguish between two tones having an intensity ratio of about $\frac{1}{4}$ db. This amounts to a 6 per cent differential, about the same as for the eye.

The value of the decibel in the size language of science lies in the enormous range of ratios that can be handled with numbers of convenient size. The ear is responsive to sounds differing in intensity by 1 million million (1,000,000,000,000) times; yet the decibel equivalent is a convenient 120 db. We have already seen that the ratio $1.06 = \frac{1}{4}$ db, a not inconveniently small number.

Although the decibel originated in the study of sound, its use has been extended to many other scientific and technological disciplines. Whenever a scientific worker must deal with ratios, the decibel becomes an indispensable tool.

What is differential calculus?
Before jumping headlong into the depths of differential calculus, it might be advisable to test its temperature by judiciously

wetting the big toe first. Calculus is the mathematics of continuous change. We see evidence all about us of continually varying phenomena. Birds fly this way and that, changing direction, speed, and height almost imperceptibly from one instant to the next. The wind, the motion of automobiles, the daily temperature, and atmospheric pressure—all these and countless other things are in an eternal state of flux. In the words of the Greeks, "All things flow."

Unless mathematics can come to grips with the elusiveness of change, scientists are powerless to understand nature in its most characteristic state. How can this be done?

At the risk of heresy, the description to follow will depart from the usual approach of textbooks on the calculus. An attempt will be made to provide an intuitively satisfying explanation of differential calculus. The only justification for such an approach is that it's simple, it works, and it tends to emphasize the uses to which the concepts may be put.

Let's consider once more the situation of a freely falling body. We have noted elsewhere that Galileo and his students had deduced the equation that explains this kind of motion—$S = \frac{1}{2}at^2$—where S is the distance fallen in t seconds and a is the acceleration due to gravity, 32.2 feet per second per second.

The next step is to clarify our thinking by drawing a *graph* of the *information* contained in the equation. At this point we must keep in mind the fact that a picture of the *trajectory* of the object would be a straight vertical line of little practical interest. What we are after is a graph that represents pictorially the relation between the distance fallen and the elapsed time at *any instant* of time.

This information is presented in Figure 51. At the end of 1 second the object will have fallen $\frac{1}{2} \times 32.2 \times 1^2 = 16.1$ feet; at the end of 2 seconds, $\frac{1}{4} \times 32.2 \times 2^2 = 64.4$ feet, and so on. (If graphs of this sort are unfamiliar, it would be well to review the question on coordinate geometry, page 55.)

Having drawn this graph, suppose we want to find the speed of the object at a specified instant—for example, after 2 seconds

273

of free fall. Without attempting to be too precise, we might reason as follows: At the end of 2 seconds, the object will have fallen 64.4 feet; and at the end of 3 seconds, it will have fallen 144.9 feet. A change in time equal to 1 second (the *dt* in the sketch)

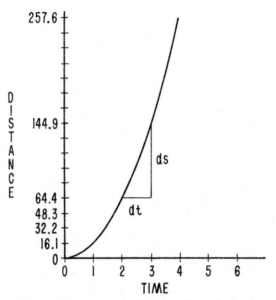

FIGURE 51. A graph showing the relation between time and distance for a falling object.

corresponds to a change in distance of 144.9 − 64.4 = 80.5 feet (the *ds* in the sketch). In other words, in the second of time following the instant we are interested in, the object falls 80.5 feet. Its average speed *during that second of time* is then $\dfrac{80.5 \text{ feet}}{1 \text{ second}} = 80.5$ feet per second.

The only difficulty with this approach is the relatively long interval of time over which the speed was computed. If we can make the time interval *dt* and the distance interval *ds very much smaller,* our answer would be much more accurate.

Let's allow the symbol *dt* to represent an extremely small difference in time—so small that it is almost, but not quite, equal to zero. Similarly, let's allow ds to represent the correspondingly small distance traveled in the time interval dt. The symbols dt and ds are called *differentitals*—hence the use of the prefix d. The

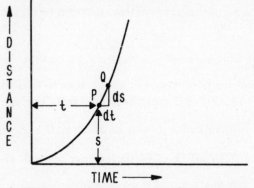

FIGURE 52. Applying the principles of differential calculus to a freely falling body.

velocity that we are after will then be $\dfrac{ds}{dt}$. We can get a general

expression for $\dfrac{ds}{dt}$ with the help of Figure 52. Let P represent any

point on the graph. P then corresponds to a distance s through which the object has fallen in a time equal to t seconds. In mathematical shorthand,

$$s = \tfrac{1}{2}at^2$$

After an extremely short length of time dt, the object will have moved a short distance ds, and will have arrived at point Q. Counting from the point of origin, the *new* time will be $t + dt$ and the corresponding distance will be $s + ds$. Since this time and distance must *satisfy* the original equation we can substitute $t + dt$ for t and $s + ds$ for s, which gives

$$s + ds = \tfrac{1}{2}a(t + dt)^2$$

or

$$s + ds = \tfrac{1}{2}a[t^2 + 2t(dt) + (dt)^2]$$

Finally,

$$s + ds = \tfrac{1}{2}at^2 + at(dt) + \tfrac{1}{2}a(dt)^2$$

275

The next step is to subtract the original equation from the last.

$$s + ds = \tfrac{1}{2}at^2 + at(dt) + \tfrac{1}{2}a(dt)^2$$
$$\underline{-s \qquad = -\tfrac{1}{2}at^2}$$
$$ds = \qquad at(dt) + \tfrac{1}{2}a(dt)^2$$

If we divide all of the terms of this equation by dt, we get

$$\frac{ds}{dt} = at + \tfrac{1}{2}a\,dt$$

Our original purpose, you will recall, was to find the speed of an object at the end of 2 seconds of free fall. The speed is now given by $\frac{ds}{dt}$ in terms of a, t, and dt. If $dt = 0.1$ seconds, $\frac{ds}{dt} = (32.2)(2) + \tfrac{1}{2}(32.2)(0.1) = 66.01$ feet per second. If $dt = 0.01$ second, $\frac{ds}{dt} = 64.561$ feet per second. Other corresponding numbers for $\frac{ds}{dt}$ and dt are tabulated below.

$\frac{ds}{dt}$, feet per second	dt, seconds
66.01	0.1
64.561	0.01
64.4161	0.001
64.40161	0.0001
64.400161	0.00001
64.4000161	0.000001
64.40000161	0.0000001

As you can see, the speed approaches 64.4 feet per second as dt is made smaller and smaller. When dt is a vanishingly small number, the term $\tfrac{1}{2}a\,dt$ may be dropped from the equation and the relation between velocity, acceleration and time becomes

$$\text{Velocity} = \frac{ds}{dt} = at$$

276

The important point here is that while ds and dt may be so small as to be negligible, their ratio $\dfrac{ds}{dt}$ does not disappear rather approaches a useful and respectable number. In the example chosen, their ratio was the velocity of a falling object. The same principles can be applied to any of the continuously changing phenomena that pervade our everyday lives.

Is the earth a sphere?

The ancient Greeks were sure that the earth was a sphere and, as we have seen, several of them had succeeded in making fairly accurate estimates of its size. For the next two thousand years or so mathematicians were limited to making greater and greater refinements in technique without seriously doubting the spherical theory. It seemed that man had gone about as far as it was possible to go in the measurement of the earth. Then the unexpected happened. A French astronomer named Richer traveled to the island of Cayenne in Guiana. Among his possessions was a very accurate pendulum clock needed in his work. He knew that the clock kept perfect time in Paris, but in Cayenne it lost $2\frac{1}{2}$ minutes of time every day! Most of us would have shortened the pendulum a bit and attributed the discrepancy to some unknown and probably unimportant accident. But not Richer. The same observation was being made by other astronomers that happened to travel toward the equator. This couldn't be coincidental—there had to be a reason and the great Isaac Newton was the man that found it. He reasoned that the pendulum depends for its rate of swing upon two factors: the length of the pendulum, and the acceleration due to gravity. In order to account for the observations, it was necessary to assume that either the pendulums grew longer as they were carried toward the equator, or the acceleration due to gravity grew less. Since the former alternative seemed highly improbable, he choose the latter. The clocks went slow because the pendulum *fell more slowly* in Cayenne than in Paris. Newton deduced that a clock must be farther from the center of the earth at the equator, so that gravity itself is less. Consequently, the earth must have an equatorial bulge and a flattening at the poles. Such a solid is called an *oblate spheroid*.

There was general agreement about the oblate spheroid until about 1700 when a father and son named Cassini threw the mathematical world into another bitter controversy. They had measured the length of an arc of the meridian through Paris and their results were in disagreement with accepted theory. The measurement showed that a degree in the northern part of the arc was shorter than a degree in the southern part. For this to be true, the earth would have to be a *prolate spheroid*—a lemon-shaped solid with the poles at each end! The battle was on. Prolate or oblate?

In 1718, the Parisian Academy of Sciences dispatched two scientific expeditions: one to Lapland and one to Peru. Each was to measure the length of 1° of latitude, so the question could be resolved once and for all. Ten years later, both expeditions had returned after bitter experiences with cold, heat, mosquitoes, hostile Indians, and internal strife. The results showed that a degree of latitude near the Arctic Circle is definitely longer than a degree at the equator. The oblate spheroid had come out the victor. When all the excitement had subsided, someone happened to recall the Cassini measurements and their line was remeasured with great care. It turned out that they had been wrong.

And so for over 200 years the earth has remained an oblate spheroid. But within the past few days a new theory has been advanced as a result of measurements made on the satellites connected with the recent international geophysical year. It seems that the earth may turn out to be pear-shaped! Will the whole question be reopened once more with all the bitterness that has characterized the past controversies? I suppose we'll have to wait and see.

Which crop-planting pattern uses available ground most economically?

In planting an orchard to use space most economically we are interested in providing a specified separation between trees while planting the greatest number in the available area. The same problem exists for corn, tomatoes, and many other crops. Fortunately, mathematics gives us a simple method of selecting the optimum arrangement.

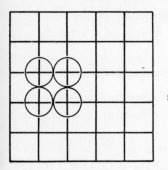

FIGURE 53. A planting pattern consisting of a mosaic of squares.

A plane surface, such as a piece of paper or a farm, can be divided into a mosaic of squares by drawing sets of horizontal and vertical lines as shown in Figure 53. A tree can be planted at each intersection and each tree will be equidistant from its

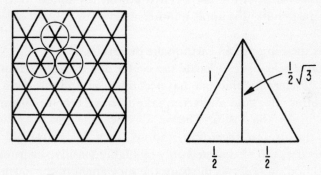

FIGURE 54. A triangular planting pattern.

neighbors. The circles indicate the area upon which a few of the trees may draw for soil, water, and sunlight. In much the same way, a plane surface can be divided into equilateral triangles or regular hexagons. Since the regular hexagon is divided by its radii into six equilateral triangles, the two arrangements are identical.

279

The circles in Figure 53 are as large as those in Figure 54, but there is not so much space lost between them.

A comparison of the two layouts will show that both have the same number of trees in a horizontal row if we neglect perimeter effects. The vertical separation between rows, however, is considerably less for the triangular arrangement. The distance from row to row is equal to the altitude of one of the equilateral triangles:

$$\text{Altitude} = \tfrac{1}{2}\sqrt{3} = 0.866 \times \text{(distance between trees)}$$

The triangular arrangement, therefore, requires only about 87 per cent as much land to set out a given number of trees as does the square plan.

If you're wondering whether other regular polygons may provide even more economical arrangements, it turns out that no other such plans are possible. The three regular polygons mentioned above are the only ones into which a plane surface is divisible. The little worker bee seems to have come to this same conclusion without the advantages of a course in Euclid. The hexagonal form of the bee's cell is admirably suited to its purposes. Just chalk this up as another marvelous activity of bees.

How is the center of an earthquake determined?

Before a scientist can locate the center of a distant earthquake, he must determine that one has occurred—and that is not a simple matter. Imagine an earthquake to have taken place somewhere under the Pacific Ocean. By the time the shock waves reach the nearest recording stations, they are extremely faint tremors that can be detected only by highly sensitive instruments. In addition, in order to measure the movement of the earth, one must have a stationary frame of reference. This is complicated by the tendency of all objects to move with the earth, as the shock waves pass by. Seismologists have solved this problem by making use of *inertia*, the tendency of a stationary body to remain at rest. If you suspend a weight from a string, the weight tends to lag behind when the support is moved. This tendency is due to inertia and it is this characteristic that forms the basis of the *seismograph*.

Locating the center of an earthquake also involves a knowledge of the way in which vibrations move from one place to another. When an earthquake occurs, the earth is caused to vibrate and the resulting impulses move away from the center of the earthquake in all directions. But there are several ways in which the earth can be made to vibrate. The two kinds of vibrations (or waves) of greatest importance in earthquake study have been given the names *P* and *S,* the letters standing for *primary* and *secondary* in order of arrival.

A *P*-wave is really a sound wave traveling through the earth. It's the fastest kind of wave set up by earthquakes. As the wave passes any point in its path the particles of the earth move back and forth in the *same* direction as the wave. *S*-waves are similar except for the direction of motion of the particles, which is at *right angles* to the direction of the wave. *P*-waves move at about 5 miles per second, while *S*-waves move at about $2\frac{3}{4}$ miles per second. This means that the *P*-wave arrives first, followed soon after by the *S*-wave. The greater the distance between the origin of the earthquake and the seismograph, the greater the time interval between the arrival of the two waves. So to determine the distance to the center of an earthquake, it's only necessary to identify the two waves on the recording device and measure the difference in arrival time.

Once the distance to the earthquake has been determined, it's possible to draw a circle on a map having this distance as its radius and the recording point as its center. Similar circles are drawn for several other distances determined at other recording stations. The point of intersection of these circles is the center of the earthquake. With certain refinements this procedure can locate the center of a distant earthquake within a very few miles.

BIBLIOGRAPHY

FOR FURTHER READING

Baker, R. H., *Introduction to Astronomy*. New York: D. Van Nostrand Company, Inc., 1947.

Bell, E. T., *The Development of Mathematics*, 2d ed. New York: McGraw-Hill Book Company, Inc., 1945.

Bell, E. T., *The Magic of Numbers*. New York: McGraw-Hill Book Company, Inc., 1946.

Bell, E. T., *Men of Mathematics*. New York: Simon and Schuster, Inc., 1937.

Bell, E. T., *Mathematics, Queen and Servant of Science*. New York: McGraw-Hill Book Company, Inc., 1941.

Collins, A. F., *Short Cuts in Figures*. New York: Edward J. Clode, 1916.

Courant, R., and Robbins, H., *What Is Mathematics?* New York: Oxford University Press, 1941.

Dantzig, T., *Number: The Language of Science*, 3d ed. New York: The Macmillan Company, 1939.

Denbow C. H., and Goedicke, V., *Foundations of Mathematics*. New York: Harper & Brothers, 1959.

Dubisch, Roy, *The Nature of Number*. New York: The Ronald Press Company, 1952.

Duncan, J. C., *Astronomy*, 5th ed. New York: Harper & Brothers, 1955.

Gamow, G., *Mr. Tompkins in Wonderland*. New York: The Macmillan Company, 1940.

Gamow, G., *One, Two, Three—Infinity*. New York: The Viking Press, Inc., 1947.

Gardner, M., *Mathematics, Magic and Mystery*. New York: Dover, 1956.

Hogben, L. T., *Mathematics for the Million*. New York: W. W. Norton & Company, 1937.

Huff, D., *How to Lie with Statistics*. New York: W. W. Norton & Company, 1954.

Kasner, E., and Newman, J., *Mathematics and the Imagination*. New York: Simon and Schuster, Inc., 1940.

Klein, Felix, *Elementary Mathematics from an Advanced Standpoint*, vol. 1. New York: The Macmillan Company, 1932.

Kline, Morris, *Mathematics in Western Culture*. New York: Oxford University Press, 1953.

Kraitchik, M., *Mathematical Recreations*, 2d edition. New York: Dover, 1953.

McKay, H., *The World of Numbers*. New York: The Macmillan Company, 1946.

Ore, Oystein, *Number Theory and Its History*. New York: McGraw-Hill Book Company, Inc., 1948.

Reid, C., *From Zero to Infinity*. New York: The Thomas Y. Crowell Company, 1955.

Titchmarsh, E. C., *Mathematics for the General Reader*. New York: Longmans, Green & Co., Inc., 1949.

APPENDIX

CONSTANTS AND CONVERSION FACTORS

	Quantity	*Abbreviation*	*Equivalent*
Length	1 millimeter	= 1 mm.	= 10^{-3} m.
	1 centimeter	= 1 cm.	= 10^{-2} m.
	1 meter	= 1 m.	= 3.2808 ft. = 39.37 in.
	1 kilometer	= 1 km.	= 10^3 m. = 0.62137 mi. = 3,281 ft.
	1 inch	= 1 in.	= 2.54 cm.
	1 foot	= 1 ft.	= 30.48 cm.
	1 mile	= 1 mi.	= 5,280 ft. = 1.60935 km.
	1 Ångström unit	= 1 Å	= 10^{-8} cm. = 10^{-6} m.
	1 micron	= 1 μ	= 10^{-4} cm.
	1 astronomical unit		= 93,005,000 mi. = 1.49678 $\times$ 10^{13} cm.
	1 light-year		= 5.880 $\times$ 10^{12} mi. = 9.463 $\times$ 10^{17} cm.
			= 63,300 astronomical units
	1 parsec		= 3.26 light-years
Weights and measures	1 cubic centimeter	= 1 cm.³	= 3.531 $\times$ 10^{-5} ft.³
	1 cubic foot	= 1 ft.³	= 1,728 in.³
	1 liter	= 1 l.	= 1,000 cm.³
	1 quart	= 1 qt.	= 1,057 cm.³
	1 gallon	= 1 gal.	= 3,785 cm.³
	1 acre		= 4.356 $\times$ 10^4 ft.²
	1 gram	= 1 g.	= 2.205 $\times$ 10^{-3} lb.
	1 kilogram	= 1 kg.	= 2.205 lb. = 10^3 g.
	1 cubic foot of water		= 62.43 lb.

1 cubic centimer of water (maximum at 3.98°C)	= 0.999973 g.
1 cubic centimeter of air (at 0°C and 760 mm.)	= 0.001293 g.

The earth

equatorial radius	= 3,963.35 mi.
polar radius	= 3,950.01 mi.
acceleration due to gravity g	= 32.17 ft./sec.^{2} = 981 cm./sec.^{2}
mass	= 6.6×10^{21} tons
1 sidereal year	= 365d 06h 09m 10s
velocity of escape	= 6.94 mi./sec.
orbital velocity	= 18.5 mi./sec. = 29.8 km./sec.
1 atmosphere	= 76 cm. of mercury = 14.7 lb./in.^{2}
solar constant	= 1.94 calories/cm.^{2}/min.

Miscellaneous

$\pi = 3.1415926536$ $\log \pi = 0.497149873$
$e = 2.7182818285$ $\log e = 0.434294482$

$\log_e 10 = 2.3026$

$\log_{10} e = 0.4343$

$\log_{10} a = \dfrac{\log_e a}{\log_e 10} = 0.4343 \log_e a$

$\log_e a = \dfrac{\log_{10} a}{\log_{10} e} = 2.3026 \log_{10} a$

1 radian = 57.2957795°
1 degree = 0.017453293 radian

velocity of light c	= 299,793 km./sec.
mass of electron m	= 9.1083×10^{-28} g.
mass of proton	= $1,836.12 \times$ mass of electron
Wien's displacement-law constant	= 0.289782 cm.-deg.
Planck's constant	= 6.62517×10^{-27} erg.-sec.
electron charge	= 1.602×10^{-19} coulomb

degrees Kelvin = K°	= 273.18 + C°
degrees Rankine = R°	= 459.72 + F°
degrees centigrade = C°	= $\frac{5}{9}$(F° − 32)
degrees Fahrenheit = F°	= $\frac{9}{5}$ × C° + 32
1 horsepower	= 0.7457 kw.
1 kilowatthour	= 3413 British thermal units (Btu)
	= 3.6 × 10⁶ joules = 3.6 × 10¹³ ergs

MENSURATION FORMULAS

Circle	r = radius
	d = diameter = $2r$
	c = circumference
	c = $2\pi r$ = πd
	area = πr^2
Sphere	Surface of sphere = $4\pi r^2 = \pi d^2$
	Volume of sphere = $\frac{4}{3}\pi r^3 = \frac{1}{6}\pi d^3$

Trigonometric Formulas

If A, B, and C are the angles of a right triangle (C the right angle) and a, b, and h the sides opposite respectively,

Sine A = sin A = $\dfrac{a}{h}$

Cosine A = cos A = $\dfrac{b}{h}$

Tangent A = tan A = $\dfrac{a}{b}$

Secant A = sec A = $\dfrac{h}{b}$

Cosecant A = csc A = $\dfrac{h}{a}$

Cotangent A = cot A = $\dfrac{b}{a}$

287

FACTORIALS AND THEIR LOGARITHMS

n	$n!$	log $n!$
1	1	0.0000
2	2	0.3010
3	6	0.7782
4	24	1.3802
5	120	2.0792
6	720	2.8573
7	5040	3.7024
8	40320	4.6055
9	362880	5.5598
10	3628800	6.5598
11	3.992×10^7	7.6012
12	4.790×10^8	8.6803
13	6.227×10^9	9.7943
14	8.718×10^{10}	10.9404
15	1.308×10^{12}	12.1166
16	2.092×10^{13}	13.3206
17	3.557×10^{14}	14.5511
18	6.402×10^{15}	15.8063
19	1.217×10^{17}	17.0853
20	2.433×10^{18}	18.3861
21	5.109×10^{19}	19.7083
22	1.124×10^{21}	21.0508
23	2.585×10^{22}	22.4125
24	6.205×10^{23}	23.7927
25	1.551×10^{25}	25.1906
26	4.033×10^{26}	26.6056
27	1.089×10^{28}	28.0370
28	3.049×10^{29}	29.4842
29	8.842×10^{30}	30.9466
30	2.653×10^{32}	32.4237
31	8.223×10^{33}	33.9150
32	2.631×10^{35}	35.4201
33	8.683×10^{36}	36.9387
34	2.952×10^{38}	38.4701

n	$n!$	$\log n!$
35	1.033×10^{40}	40.0141
36	3.720×10^{41}	41.5705
37	1.376×10^{43}	43.1386
38	5.230×10^{44}	44.7185
39	2.040×10^{46}	46.3096
40	8.159×10^{47}	47.9116
41	3.345×10^{49}	49.5244
42	1.405×10^{51}	51.1477
43	6.042×10^{52}	52.7812
44	2.658×10^{54}	54.4246
45	1.196×10^{56}	56.0777
46	5.503×10^{57}	57.7406
47	2.586×10^{59}	59.4126
48	1.241×10^{61}	61.0938
49	6.083×10^{62}	62.7841
50	3.041×10^{64}	64.4830
51	1.551×10^{66}	66.1906
52	8.066×10^{67}	67.9067
53	4.275×10^{69}	69.6309
54	2.308×10^{71}	71.3632
55	1.270×10^{73}	73.1038
56	7.110×10^{74}	74.8519
57	4.053×10^{76}	76.6078
58	2.351×10^{78}	78.3713
59	1.387×10^{80}	80.1421
60	8.321×10^{81}	81.9202

Powers and Roots

No.	Sq.	Sq. Root	Cube	Cube Root	No.	Sq.	Sq. Root	Cube	Cube Root
1	1	1.000	1	1.000	51	2,601	7.141	132,651	3.708
2	4	1.414	8	1.260	52	2,704	7.211	140,608	3.733
3	9	1.732	27	1.442	53	2,809	7.280	148,877	3.756
4	16	2.000	64	1.587	54	2,916	7.348	157,464	3.780
5	25	2.236	125	1.710	55	3,025	7.416	166,375	3.803
6	36	2.449	216	1.817	56	3,136	7.483	175,616	3.826
7	49	2.646	343	1.913	57	3,249	7.550	185,193	3.849
8	64	2.828	512	2.000	58	3,364	7.616	195,112	3.871
9	81	3.000	729	2.080	59	3,481	7.681	205,379	3.893
10	100	3.162	1,000	2.154	60	3,600	7.746	216,000	3.915
11	121	3.317	1,331	2.224	61	3,721	7.810	226,981	3.936
12	144	3.464	1,728	2.289	62	3,844	7.874	238,328	3.958
13	169	3.606	2,197	2.351	63	3,969	7.937	250,047	3.979
14	196	3.742	2,744	2.410	64	4,096	8.000	262,144	4.000
15	225	3.873	3,375	2.466	65	4,225	8.062	274,625	4.021
16	256	4.000	4,096	2.520	66	4,356	8.124	287,496	4.041
17	289	4.123	4,913	2.571	67	4,489	8.185	300,763	4.062
18	324	4.243	5,832	2.621	68	4,624	8.246	314,432	4.082
19	361	4.359	6,859	2.668	69	4,761	8.307	328,509	4.102
20	400	4.472	8,000	2.714	70	4,900	8.367	343,000	4.121
21	441	4.583	9,261	2.759	71	5,041	8.426	357,911	4.141
22	484	4.690	10,648	2.802	72	5,184	8.485	373,248	4.160
23	529	4.796	12,167	2.844	73	5,329	8.544	389,017	4.179
24	576	4.899	13,824	2.884	74	5,476	8.602	405,224	4.198
25	625	5.000	15,625	2.924	75	5,625	8.660	421,875	4.217
26	676	5.099	17,576	2.962	76	5,776	8.718	438,976	4.236
27	729	5.196	19,683	3.000	77	5,929	8.775	456,533	4.254
28	784	5.291	21,952	3.037	78	6,084	8.832	474,552	4.273
29	841	5.385	24,389	3.072	79	6,241	8.888	493,039	4.291
30	900	5.477	27,000	3.107	80	6,400	8.944	512,000	4.309
31	961	5.568	29,791	3.141	81	6,561	9.000	531,441	4.327
32	1,024	5.657	32,768	3.175	82	6,724	9.055	551,368	4.344
33	1,089	5.745	35,937	3.208	83	6,889	9.110	571,787	4.362
34	1,156	5.831	39,304	3.240	84	7,056	9.165	592,704	4.380
35	1,225	5.916	42,875	3.271	85	7,225	9.220	614,125	4.397
36	1,296	6.000	46,656	3.302	86	7,396	9.274	636,056	4.414
37	1,369	6.083	50,653	3.332	87	7,569	9.327	658,503	4.431
38	1,444	6.164	54,872	3.362	88	7,744	9.381	681,472	4.448
39	1,521	6.245	59,319	3.391	89	7,921	9.434	704,969	4.465
40	1,600	6.325	64,000	3.420	90	8,100	9.487	729,000	4.481
41	1,681	6.403	68,921	3.448	91	8,281	9.539	753,571	4.498
42	1,764	6.481	74,088	3.476	92	8,464	9.592	778,688	4.514
43	1,849	6.557	79,507	3.503	93	8,649	9.644	804,357	4.531
44	1,936	6.633	85,184	3.530	94	8,836	9.695	830,584	4.547
45	2,025	6.708	91,125	3.557	95	9,025	9.747	875,375	4.563
46	2,116	6.782	97,336	3.583	96	9,216	9.798	884,736	4.579
47	2,209	6.856	103,823	3.609	97	9,409	9.849	912,673	4.595
48	2,304	6.928	110,592	3.634	98	9,604	9.899	941,192	4.610
49	2,401	7.000	117,649	3.659	99	9,801	9.950	970,299	4.626
50	2,500	7.071	125,000	3.684	100	10,000	10.000	1,000,000	4.642

Common Logarithms

N	0	1	2	3	4	5	6	7	8	9
10	0000	0043	0086	0128	0170	0212	0253	0294	0334	0374
11	0414	0453	0492	0531	0569	0607	0645	0682	0719	0755
12	0792	0828	0864	0899	0934	0969	1004	1038	1072	1106
13	1139	1173	1206	1239	1271	1303	1335	1367	1399	1430
14	1461	1492	1523	1553	1584	1614	1644	1673	1703	1732
15	1761	1790	1818	1847	1875	1903	1931	1959	1987	2014
16	2041	2068	2095	2122	2148	2175	2201	2227	2253	2279
17	2304	2330	2355	2380	2405	2430	2455	2480	2504	2529
18	2553	2577	2601	2625	2648	2672	2695	2718	2742	2765
19	2788	2810	2833	2856	2878	2900	2923	2945	2967	2989
20	3010	3032	3054	3075	3096	3118	3139	3160	3181	3201
21	3222	3243	3263	3284	3304	3324	3345	3365	3385	3404
22	3424	3444	3464	3483	3502	3522	3541	3560	3579	3598
23	3617	3636	3655	3674	3692	3711	3729	3747	3766	3784
24	3802	3820	3838	3856	3874	3892	3909	3927	3945	3962
25	3979	3997	4014	4031	4048	4065	4082	4099	4116	4133
26	4150	4166	4183	4200	4216	4232	4249	4265	4281	4298
27	4314	4330	4346	4362	4378	4393	4409	4425	4440	4456
28	4472	4487	4502	4518	4533	4548	4564	4579	4594	4609
29	4624	4639	4654	4669	4683	4698	4713	4728	4742	4757
30	4771	4786	4800	4814	4829	4843	4857	4871	4886	4900
31	4914	4928	4942	4955	4969	4983	4997	5011	5024	5038
32	5051	5065	5079	5092	5105	5119	5132	5145	5159	5172
33	5185	5198	5211	5224	5237	5250	5263	5276	5289	5302
34	5315	5328	5340	5353	5366	5378	5391	5403	5416	5428
35	5441	5453	5465	5478	5490	5502	5514	5527	5539	5551
36	5563	5575	5587	5599	5611	5623	5635	5647	5658	5670
37	5682	5694	5705	5717	5729	5740	5752	5763	5775	5786
38	5798	5809	5821	5832	5843	5855	5866	5877	5888	5899
39	5911	5922	5933	5944	5955	5966	5977	5988	5999	6010
40	6021	6031	6042	6053	6064	6075	6085	6096	6107	6117
41	6128	6138	6149	6160	6170	6180	6191	6201	6212	6222
42	6232	6243	6253	6263	6274	6284	6294	6304	6314	6325
43	6335	6345	6355	6365	6375	6385	6395	6405	6415	6425
44	6435	6444	6454	6464	6474	6484	6493	6503	6513	6522
45	6532	6542	6551	6561	6571	6580	6590	6599	6609	6618
46	6628	6637	6646	6656	6665	6675	6684	6693	6702	6712
47	6721	6730	6739	6749	6758	6767	6776	6785	6694	6803
48	6812	6821	6830	6839	6848	6857	6866	6875	6884	6893
49	6902	6911	6920	6928	6937	6946	6955	6964	6972	6981
50	6990	6998	7007	7016	7024	7033	7042	7050	7059	7067
51	7076	7084	7093	7101	7110	7118	7126	7135	7143	7152
52	7160	7168	7177	7185	7193	7202	7210	7218	7226	7235
53	7243	7251	7259	7267	7275	7284	7292	7300	7308	7316
54	7324	7332	7340	7348	7356	7364	7372	7380	7388	7396

Common Logarithms

N	0	1	2	3	4	5	6	7	8	9
55	7404	7412	7419	7427	7435	7443	7451	7459	7466	7474
56	7482	7490	7497	7505	7513	7520	7528	7536	7543	7551
57	7559	7566	7574	7582	7589	7597	7604	7612	7619	7627
58	7634	7642	7649	7657	7664	7672	7679	7686	7694	7701
59	7709	7716	7723	7731	7738	7745	7752	7760	7767	7774
60	7782	7789	7796	7803	7810	7818	7825	7832	7839	7846
61	7853	7860	7868	7875	7882	7889	7896	7903	7910	7917
62	7924	7931	7938	7945	7952	7959	7966	7973	7980	7987
63	7993	8000	8007	8014	8021	8028	8035	8041	8048	8055
64	8062	8069	8075	8082	8089	8096	8102	8109	8116	8122
65	8129	8136	8142	8149	8156	8162	8169	8176	8182	8189
66	8195	8202	8209	8215	8222	8228	8235	8241	8248	8254
67	8261	8267	8274	8280	8287	8293	8299	8306	8312	8319
68	8325	8331	8338	8344	8351	8357	8363	8370	8376	8382
69	8388	8395	8401	8407	8414	8420	8426	8432	8439	8445
70	8451	8457	8463	8470	8476	8482	8488	8494	8500	8506
71	8513	8519	8525	8531	8537	8543	8549	8555	8561	8567
72	8573	8579	8585	8591	8597	8603	8609	8615	8621	8627
73	8633	8639	8645	8651	8657	8663	8669	8675	8681	8686
74	8692	8698	8704	8710	8716	8722	8727	8733	8739	8745
75	8751	8756	8762	8768	8774	8779	8785	8791	8797	8802
76	8808	8814	8820	8825	8831	8837	8842	8848	8854	8859
77	8865	8871	8876	8882	8887	8893	8899	8904	8910	8915
78	8921	8927	8932	8938	8943	8949	8954	8960	8965	8971
79	8976	8982	8987	8993	8998	9004	9009	9015	9020	9025
80	9031	9036	9042	9047	9053	9058	9063	9069	9074	9079
81	9085	9090	9096	9101	9106	9112	9117	9122	9128	9133
82	9138	9143	9149	9154	9159	9165	9170	9175	9180	9186
83	9191	9196	9201	9206	9212	9217	9222	9227	9232	9238
84	9243	9248	9253	9258	9263	9269	9274	9279	9284	9289
85	9294	9299	9304	9309	9315	9320	9325	9330	9335	9340
86	9345	9350	9355	9360	9365	9370	9375	9380	9385	9390
87	9395	9400	9405	9410	9415	9420	9425	9430	9435	9440
88	9445	9450	9455	9460	9465	9469	9474	9479	9484	9489
89	9494	9499	9504	9509	9513	9518	9523	9528	9533	9538
90	9542	9547	9552	9557	9562	9566	9571	9576	9581	9586
91	9590	9595	9600	9605	9609	9614	9619	9624	9628	9633
92	9638	9643	9647	9652	9657	9661	9666	9671	9675	9680
93	9685	9689	9694	9699	9703	9708	9713	9717	9722	9727
94	9731	9736	9741	9745	9750	9754	9759	9763	9768	9773
95	9777	9782	9786	9791	9795	9800	9805	9809	9814	9818
96	9823	9827	9832	9836	9841	9845	9850	9854	9859	9863
97	9868	9872	9877	9881	9886	9890	9894	9899	9903	9908
98	9912	9917	9921	9926	9930	9934	9939	9943	9948	9952
99	9956	9961	9965	9969	9974	9978	9983	9987	9991	9996

Trigonometric Functions

Deg.	Rad.	Sin	Cos	Tan	Cot		
0	0.0000	0.0000	1.0000	0.0000		1.5708	90
1	0.0175	0.0175	0.9998	0.0175	57.290	1.5533	89
2	0.0349	0.0349	0.9994	0.0349	28.636	1.5359	88
3	0.0524	0.0523	0.9986	0.0524	19.081	1.5184	87
4	0.0698	0.0698	0.9976	0.0699	14.301	1.5010	86
5	0.0873	0.0872	0.9962	0.0875	11.430	1.4835	85
6	0.1047	0.1045	0.9945	0.1051	9.5144	1.4661	84
7	0.1222	0.1219	0.9925	0.1228	8.1443	1.4486	83
8	0.1396	0.1392	0.9903	0.1405	7.1154	1.4312	82
9	0.1571	0.1564	0.9877	0.1584	6.3138	1.4137	81
10	0.1745	0.1736	0.9848	0.1763	5.6713	1.3963	80
11	0.1920	0.1908	0.9816	0.1944	5.1446	1.3788	79
12	0.2094	0.2079	0.9781	0.2126	4.7046	1.3614	78
13	0.2269	0.2250	0.9744	0.2309	4.3315	1.3439	77
14	0.2443	0.2419	0.9703	0.2493	4.0108	1.3265	76
15	0.2618	0.2588	0.9659	0.2679	3.7321	1.3090	75
16	0.2793	0.2756	0.9613	0.2867	3.4874	1.2915	74
17	0.2967	0.2924	0.9563	0.3057	3.2709	1.2741	73
18	0.3142	0.3090	0.9511	0.3249	3.0777	1.2566	72
19	0.3316	0.3256	0.9455	0.3443	2.9042	1.2392	71
20	0.3491	0.3420	0.9397	0.3640	2.7475	1.2217	70
21	0.3665	0.3584	0.9336	0.3839	2.6051	1.2043	69
22	0.3840	0.3746	0.9272	0.4040	2.4751	1.1868	68
23	0.4014	0.3907	0.9205	0.4245	2.3559	1.1694	67
24	0.4189	0.4067	0.9135	0.4452	2.2460	1.1519	66
25	0.4363	0.4226	0.9063	0.4663	2.1445	1.1345	65
26	0.4538	0.4384	0.8988	0.4877	2.0503	1.1170	64
27	0.4712	0.4540	0.8910	0.5095	1.9626	1.0996	63
28	0.4887	0.4695	0.8829	0.5317	1.8807	1.0821	62
29	0.5061	0.4848	0.8746	0.5543	1.1434	1.0647	61
30	0.5236	0.5000	0.8660	0.5774	1.7321	1.0472	60
31	0.5411	0.5150	0.8572	0.6009	1.6643	1.0297	59
32	0.5585	0.5299	0.8480	0.6249	1.6003	1.0123	58
33	0.5760	0.5446	0.8387	0.6494	1.5399	0.9948	57
34	0.5934	0.5592	0.8290	0.6745	1.4826	0.9774	56
35	0.6109	0.5736	0.8192	0.7002	1.4281	0.9599	55
36	0.6283	0.5878	0.8090	0.7265	1.3764	0.9425	54
37	0.6458	0.6018	0.7986	0.7536	1.3270	0.9250	53
38	0.6632	0.6157	0.7880	0.7813	1.2799	0.9076	52
39	0.6807	0.6293	0.7771	0.8098	1.2349	0.8901	51
40	0.6981	0.6428	0.7660	0.8391	1.1918	0.8727	50
41	0.7156	0.6561	0.7547	0.8693	1.1504	0.8552	49
42	0.7330	0.6691	0.7431	0.9004	1.1106	0.8378	48
43	0.7505	0.6820	0.7314	0.9325	1.0724	0.8203	47
44	0.7679	0.6947	0.7193	0.9657	1.0355	0.8029	46
45	0.7854	0.7071	0.7071	1.0000	1.0000	0.7854	45
		Cos	Sin	Cot	Tan	Rad.	Deg.

INDEX

Light pipe, 167
Limits, 170
Line, straight, equation of, 59
Lobachevsky, Nikolai, 16
Logarithms, 246
  and decibels, 271
Longitude, length of one degree, 134
Loudness, 269
Loxodromic curves, 132

Magellanic Clouds, 232
Magic tricks, 4, 9, 99
Magnitude, of celestial bodies, 210, 221
Mantissa, 246
Map-making, 71
  colors required for, 130
Markup, 109
Mass, and energy, 118
  center of, 242
Mass-ratio, 251
Mathematical induction, 202
Mathematics, applied, 54
Mean, 238
Median, 238
Mercator projection, 71
Mercury, 209
Metals, expansion of, 175
Meteorology, 117
Meteors, origin of, 180
  velocity of, 180, 182
Meteor showers, 178
Meteor trails, communication via, 185
Michelson, A. A., 28
Mil, 256
Mile, nautical, 200
Möbius, August Ferdinand, 3
Mode, 238
Model, physical, 49
Momentum, 47
Month, sidereal, 53
  synodic, 53
Moon, and the tides, 66
  eclipse of, 72
  effect of, on orbit of earth, 242
  first measurement of distance to, 77

Moon (*cont.*)
  gravity on, 89
  height of mountains of, 133
  lack of atmosphere, 88
  passing over a star, 53
  speed of, 264
Moonlight, causing stars to disappear, 271
Mountains, lunar, height of, 133
Morley, E. W., 28
Multiplication, by logarithms, 246
  on one's fingers, 63
  peasant, 158
  short cuts in, 143
Musical scale, evolution of, 171

Navigation, celestial, 200
Negatives, two make a positive, 136
Neptune, discovery of, 167
Newton, Sir Issac, 103, 277
Node of emission, 101
Non-Euclidean geometry, 16
  and relativity, 29
North, ancient determination of, 45
North Star, location of, 196
  movement of, 194
Number, most important, 234
Number system, 234, 244
Numbers, amicable, 143
  cardinal, 61
  complex, 128
  continuous, 259
  directed, 126
  hidden relationships in, 197
  imaginary, 126, 149
  irrational, 257
  large, 114
  ordinal, 61
  perfect, 142
  rational, 31
Number theory, 31, 35, 49, 194, 197, 202, 209, 222, 234, 244, 257

Ohm's law, 20
Operative signs, 124
Optics, 167

300